Sustainable Water Technologies

Sustainable Water Technologies

Editor

Dalmiro Martell

scitus
academics

Sustainable Water Technologies
Edited by **Dalmiro Martell**

ISBN: 978-1-68117-711-3
Library of Congress Control Number: 2016958747

SCITUS Academics LLC,
616, Corporate Way, Suite 2, 4766,
Valley Cottage, NY 10989

www.scitusacademics.com

Notice

Printed in United States of America on Acid Free Paper

Preface

On this Earth, drinking water is one of the best resources. People couldn't survive without it. Nevertheless, we all use it too often as an unlimited resource that never ends. Emerging pressure from economic growth and population on water resources, climate change, pollution and other challenges has a major impact on our environmental, economic and social security. Many of our most important aquifers are over pumped, which affects widespread declines in groundwater. Management of water resources, sustainability of water resources, quality and quantity of ground and surface water, reuse and use of water, interaction between surface and ground water, aquifer recharge, storage and more are addressed by sustainable water technologies. In addition to sustainable products and processes to protect human health and the environment, save energy and water and reduce material use, the emphasis is placed on emerging applications in nanotechnology, biotechnology and info technology.

This book aims to bring together industry and academia practitioners and researchers to focus on the development of advanced technologies, a key component in overcoming the ongoing water crisis. It emphasizes emerging technologies and strategies that will improve the sustainability of water for future generations. The scope of sustainable water resources management is the responsibility of water managers and users to ensure that water resources are equitably and efficiently allocated and used to achieve social, economic and environmental benefits. Future policies for the sustainable management of water resources will be heavily influenced by climate change and climate variability, including problems of seasonality, such as changes in the timing of annual rainfall patterns or snow pack melt periods. Recognition of the complexity and variety of water resources management is essential from a policy perspective, as it implies that there is no one-size-fits-all policy solution to improve the management of water resources. Water resource management policies must be customized and focused on specific situations for regions and countries within countries. This reflects the wide variety of water basins from the region to international levels in terms of: heterogeneity of water sources; links between water resources and water pollution issues; allocation of water between consumption uses, such as agriculture, power generation, industry, households and green needs; and management of the complex institute. Innovative technology promises to turn waste water into a source and energy source for drinking water.

Table of Contents

1

CHAPTER

Rural Water Supply Management and Sustainability: The Case of Adama Area, Ethiopia

Pieter Desnerck*, Janet M. Lees , Chris T. Morley

Department of Engineering, University of Cambridge, Trumpington Street, CB2 1PZ Cambridge, United Kingdom

ABSTRACT

Wise utilization of water resources is becoming very important as world faces water crises. The main objective of this study was to investigate the rural water supply systems with case study in Adama area, in central Ethiopia. Both quantitative and qualitative data were collected and analyzed. Four sample water schemes were selected and totally 148 (63 were female) representative households were selected for answering the questionnaires. Key informant interviews and group discussions were also conducted. The study assessed issues such as community participation, water committee empowerment, management and governance of water supply schemes, women participation, functional status of water supply scheme, sanitation and hygiene issues, external support, and monitoring system of water supply schemes. The findings indicated that the community participation in planning and implementation was very good while monitoring mechanism of operation and management as well as community participation on choice of technology was poor. The water schemes were located at reasonable distances i.e. less than 2 km in most cases and the time taken for round trip to fetch water from source was less than or equal to 30 minutes in most cases, however the queuing time was more than an hour. The water supply was inadequate as only about 15% of beneficiaries could get 20 liters of water per day per capita. The water sources were exposed in many cases to human waste, wild life, livestock and uncontrolled flooding. Sanitary practices in the study area were poor as only about 3.4% had ventilated and improved pit latrine and open pit and/or open field defecation were widely practiced.

Keywords: Ethiopia; Rural Water Supply; Water Fee; Community Participation; Safe Drinking Water

1. INTRODUCTION

Wise utilization of water resources is becoming very important as world faces water crises which could hold back human development. According to Millennium Development Goal (MDG) summit Report [1], progress on the MDG 7 target "to reduce by half the proportion of people without sustainable access to safe drinking water and basic sanitation by 2015" is presently on pathway. However rural areas in developing countries across the world remain severely underprivileged, with eight out of ten people not having access to safe water supply. As per WHO and UNICEF [2], 87% of the world population could have access to safe drinking water, a progress of 10% within the last two decades. The report also pointed out that about 884 million people worldwide, out of which 37% living in sub-Saharan Africa, still utilize drinking water from unsafe supply spots. Africa is lagging behind the attainment of the MDG as 340 million Africans lack access to safe drinking water [3]. The proportion of the African population who had access to safe drinking water accounted for only 60% by 2010, which is about 11% increase compared to the situation in 1990 [2]. In Ethiopia, the percentage was about 68.5% by 2010 indicating good efforts made in the sector during the last two decades [4].

As per urban-rural disparities concerning access to safe drinking water, out of the world population who lacked access to safe drinking water, about 84% were living in rural areas [5]. Accordingly, urban safe drinking water coverage for Africa was estimated to be 85% (281 million people) while the rural coverage was about 51% (294 million people) by 2008 [6]. In this respect, Ethiopia has made an encouraging progress as access to safe drinking water has increased from 35% and 80% in 2005 to 65.8% and 91.5% in 2010 for rural and urban areas respectively [1].

About 84% of Ethiopian populations live in rural areas. For sizable proportion of the rural population, the major sources of drinking water are unprotected springs, ponds, rivers, and hand dug wells which are exposed to contamination caused by human beings, livestock, wildlife and uncontrolled flooding. The safety and quality of drinking water is further in jeopardy as the culture of open defecation has been socially accepted and widely practiced in most of the rural settings and partly in urban areas as well [7]. As a result, the prevalence of water borne diseases has increased at alarming rate. As noted by Dessalegn [8], rural safe drinking water supply provision has commenced in Ethiopia during the late 1950s. Since then, efforts have been made to provide safe drinking water for rural areas although there was more focus on urban water supply before two decades. This condition coupled with other factors, has contributed a lot for the low level of safe drinking water supply provisions of the country until quite recently. Access to safe drinking water has been improved from 19% in 1990 [1] to 68.5% by 2010 [4].

The driving force behind the expansion of access to safe drinking water in Ethiopia was attributed to the incidence of drought and famine in the 70s and the 80s. In response to this devastating situation, and adverse effects associated with years of environmental crises, quite a lot of multi-lateral and bilateral international NGOs, donor agencies and indigenous organizations have devoted significant proportion of their fund for the provision of rural safe drinking water supply and vigorously engaged in this endeavors [8]. Despite active mobilization of resources by international and local NGOs and the Ethiopian government, the national safe drinking water coverage of the country has not been improved that much especially in the rural areas [9]. The main reasons for this very low level of performance in the supply of safe drinking water, and the quandary for not efficiently utilizing the water resources potential of the country towards realizing sustainable water supply, is attributed to lack of articulate and holistic water policy and insufficient investment for safe drinking water supply [10]. In addition to this, communities lack capability in managing the water supply schemes. Such problems indicate the need of in depth studies to investigate the rural water supply systems in the country and finding out the strategies to improve the sustainability of the schemes.

The objective of this study was to investigate the rural water supply systems in Adama area of central Ethiopia. The study focused on assessing household water use practices, assessing communities' attitudes towards water safety and benefits of safe water supply, investigating the contributions of community for water source protection and maintenance, and investigating institutional approaches to enhance the sustainability of water supply schemes. This contributes towards the knowledge base for future research and development of water supply systems in the area. Understanding these aspects of rural water supply systems can give an insight into developing a useful strategy that can potentially address water supply system with lasting benefits over time for the intended communities.

2. SUSTAINABILITY OF WATER SUPPLY SCHEMES

The concept of sustainability has taken its root from the debate on sustainable development during the early 70s. It becomes a concept that is found out to be more "complex and contested" [11]. As per the United Nations document entitled "Our Common Future" (1987) "sustainable development is development that meets the needs of the present generations without compromising the ability of future generations to meet their own needs." This being the case, different organizations used their own version of definition to address their intended objective.

Accordingly, various studies conducted pertaining to water supply services have produced definitions concerning sustainability in the context of water supply projects. Most of these definitions capitalize on financing of regular operation and maintenance costs by users, minimal external assistance in the long term, and continued flow of benefits over a long period [12]. In this study, the following water supply sustainability definition of Abrams [13] has been adapted: "Sustainability is about whether or not water and sanitation services and good hygiene practices continue to work over time". According to this definition, the achievement of sustainability engrosses the realization of enduring "beneficial" changes in rural water services. In this case, the issue of sustainability is considered further than limiting itself on technical functionality debate; the expression "beneficial" highlights the outcome on the lives of people and it indicates to services other than technology [14]. Over years, several conceptual frameworks have been produced to better understand the essence of rural water supply sustainability. Among those developed conceptualization frameworks, the one that has been shared by many researchers has five key dimensions [12,13]: institutional (organizational), social, environmental, technical, and financial. It is well noted that the success of lasting sustainable water supply services is dependent on the interaction of a combination of factors that give due emphasis for community participation, external collaboration and technical support in order to ensure operation and maintenance of the system [12,13]. In order to have a close look at the interplay of these factors, the recently developed conceptual framework of sustainable rural water services [15] is presented in **Figure 1**.

Factors listed above are interdependent, interactive and crucial for achieving sustainable water services with a corresponding behavior changes over time [16]. This delicate balance of interacting factors requires

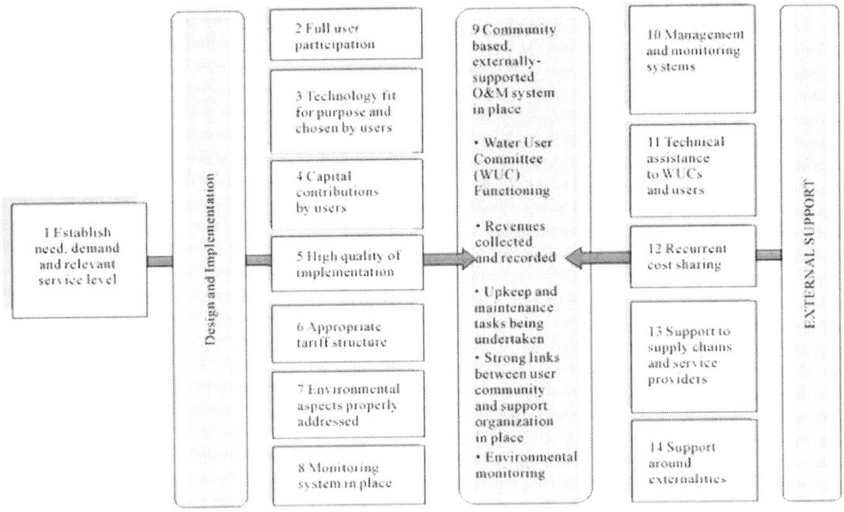

Figure 1. Conceptual framework for sustainable rural water supply services [15].

the involvement of a number of stakeholders working together. Sustainability will be achieved only if all stakeholders including communities, Non-Governmental Organizations (NGOs), and governmental offices at different levels and the private sector have sufficient capacity and initiative to do their role [17]. In this case community members are called for to make informed choices regarding participation in the project, willingness to share project cost and commitment to bear associated contribution.

Full community participation promotes a proactive process in which the beneficiaries influence the development and management of development projects rather than merely receiving a share of project benefits. Community participation creates an enabling environment for sustainability by allowing users, as a group, to select the level of services for which they are willing to pay, to guide key investment and management decisions, and also to make choices and commit resources in support of these choices [18]. The kind of technology that cannot serve the best interest of the beneficiary in terms of the quality of installation and cause further problem for maintenance could negatively impact the sustainability of water supply projects. Technology that fits for purpose of the project and chosen by users needs to be in place [16].

3. METHODOLOGY

3.1. The Study Area

Define Adama district is located in Central Ethiopia, in Oromia Region state (see **Figure 2**). The topography of the district lies within 1500 - 2300 m above sea level and is dominated by surging plains that involve extensive ridges all along its western boundaries. Significant proportion of the district is situated in a sub-tropical agroclimatic zone [19]. Adama district is one of highly populated districts. According to population and housing census conducted in 2007 by central statistics agency (CSA), the total population of Adama district is 155,321 as of 2007 [20]. The urban population accounts for 16.9%, whereas the rural population accounts for 83.1% of the total population. Concerning safe drinking water access of the district, it was reported that the level of coverage for rural area is 24% and coverage for total the population of the district is 69% [19].

3.2. Research Design, Sampling Procedures and Method of Data Collection

The study has involved both quantitative & qualitative research methods to assess the contemporary situations of water supply services. The study has emphasized on characteristic feature of selected water supply schemes and appropriate investigative analysis of observed incidents and factors as per their relationships to the context. Out of the existing 27 water supply schemes in Adama district, 4 water schemes constructed during the last 15 years were selected based on the type of technology used, management practices and system of operation. To this effect, the 4 water supply schemes with a total of 15 water distribution points/public water taps were identified as unit of analysis for this research. These water supply schemes are located at 4 sample communities namely, Adullala Hate Aroreti, Bubissa Kussaye, Cheka Dewero and Geldia Galiye (see **Figure 2**).

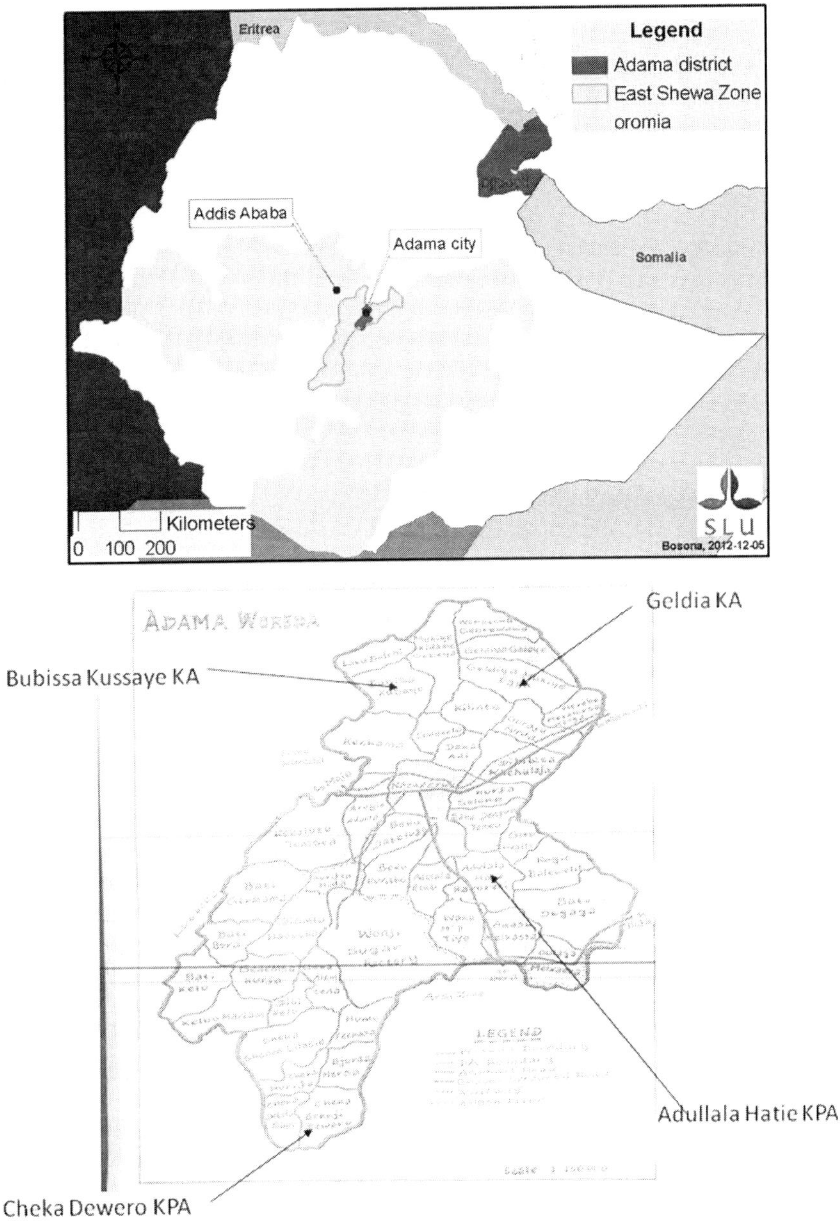

Source: Adama district finance & economic development office

Figure 2. Map of study area: Location of Adama district in Oromia region (the upper figure) and the location of sample communities (the lower figure). KA—kebele association; KPA—kebele peasant association; Kebele is administration level accountable to district level.

The four schemes were selected based on type of water sources and operational mechanisms. Adullala Hate Aroreti water supply scheme consists of boreholes operated by diesel generator, while Geldia Galiye water supply scheme consists of boreholes operated by hydropower. Cheka Dewero water supply scheme is gravityflown-pipe water supply services, while Bubissa Kussaye water supply scheme consists of ponds and hand pumps.

The sample size for each community was extracted from list of beneficiaries who can access the water supply points within the range of 1.5 kilometer radius based on the list of water supply schemes inventory of Adama district. Accordingly, 49, 15, 54 and 30 households were selected randomly from Adullala Hate, Bubissa Kussaye, Cheka Dewero and Geldia Galiye communities respectively. These 148 households were used as source of primary data for this study. Respondents are picked systematically so that all beneficiaries of the selected water points are fairly represented. About 50% of the respondents (representing the households) were illiterate. The gender and age distribution of the respondents are presented in **Table 1**.

Four focus group discussions (FGD) have been conducted, one FGD in each sample community. In each FGD 10 community members (elders, men and women) participated. Participants pertaining to key informant (KI) interview were selected from community members who are in one form or another have participated in mobilizing the community towards the realization of the water supply schemes under study. Accordingly, key informant interviews that involved 4 participants from each of the four sample communities were conducted.

Before starting detailed data collection, some general information pertaining to the socio-economic, demographic and physical characteristics, settlement patterns, and water supply schemes of the communities under study were gathered. This information has been used as a base for planning the field data collection and determining the selection of the sample population. Based on this information, different data collection instruments (survey questionnaires, interview guides and guiding questions for focus group discussions) were prepared and checked through consulting with experts and conducting initial interviews to obtain feedbacks for pre-testing. The feedbacks were analyzed and the necessary adjustment and corrections has been effected on the interview guides, survey questionnaires and guiding questions. These improved data collection instruments were used to conduct the actual data collection. To this effect, household survey that involves sample respondents has been administered and relevant quantitative data was gathered from the field. By employing participatory rural appraisal (PRA) techniques, pertinent qualitative data was gathered from focus group discussions (FGD). Interview sessions with key informants (KI) have been conducted involving

Table 1. Gender and age range of the respondents.

Gender	Age range of the respondents				Total
	15 - 30	31 - 45	46 - 60	>60	
Male	20	36	23	6	85
Female	18	23	16	6	63
Total	38	59	39	12	148

community influential's, elders and others who are active in community mobilization. In addition to pre-testing efforts, triangulation method has been used to ensure the quality and reliability of the data. For this, available documents and technical reports were used. The findings are presented and discussed in the following sections.

4. MAJOR FINDINGS AND DISCUSSIONS

4.1. Summary of Major Findings from FGD and KI Interview Sessions

Table 2 presents summary of response (consolidated responses) given under each issue of FGD. Ratings 0, 1, 2, 3, 4 represent given consolidated responses consented by each FGD. For each issue discussed by FGD, the score value is given in the last column (scoring column) for each of the four sample communities. For example, for the issue of community participation in planning and implementation of the water supply projects the score values are 4, 4, 3, 4 indicating that the community participation is good (score value = 3) in one sample community and very good (score value = 4) in three sample communities. The results of FGD analysis indicate that the rate of community participation in planning and implementation is very good while monitoring mechanism of operation and management as well as community participation on choice of technology is poor.

Similarly, the consolidated responses from KI interviews sessions are presented in **Table 3**. For instance, in three of the schemes the water fee paid by beneficiaries is very sufficient for operation and maintenance (score value = 4) while in one scheme, it is not sufficient (score value = 0) as there is no access to safe drinking water yet (this scheme is not treated, unprotected well and exposed for contamination). **Table 3** also points out that community participation in planning and implementation of water supply schemes is very good while collection and control mechanisms of water fee as well as mechanisms of monitoring the operation and management of the schemes are poor. Most of the issues indicated in these two tables (Tables 2 and 3) are also discussed in detail in the following sections.

4.2. Access to Safe Drinking Water and Water Use Practices by Communities

In this study, 65% of the total respondents have stated that they use water from improved sources for drinking, cooking and sanitation. About 23% of the respondents have pointed out that women are responsible and sole supplier of water for household requirements. Studies by WHO and UNICEF [2] indicated that families who don't have pipe born water connected to their premises usually assign women in the household to fetch water for house hold consumption. An enhanced access to safe drinking water provides the deprived, especially women, an opportunity to be in command of vital aspects of their livelihood and maximize their sense of confidence/self-esteem [21]. Access to safe drinking water and improved sanitation is fundamental and placed at the very center of concerted action towards poverty reduction.

Regarding the travel time taken to collect water from the source 23% of the sample population have responded as less than or equal to 15 minutes while 41% have responded as 16 - 30 minutes. Respondents representing 17% of the sample population have mentioned 31 - 45 minutes while 14% responded 46 - 60 minutes. On aggregate, 64% of the respondents have responded that they spent less than or equal to 30 minutes to fetch water (round trip). In this respect, previous studies [2] indicated that household members who spent more than 30 minutes to collect water usually fail to fulfill the house hold daily water consumption. The time spent due to multiple trips to collect water is so high in economic terms and contributes to lower productivity. According to WHO and UNICEF [2], sizable proportion of the population in many Sub-Saharan Africa countries, devote more than 30 minutes round trip to fetch water from communal collection points, to be used for house hold consumption. In rural Ethiopian context, residents in rural areas travel more than an hour in order to fetch unsafe water from unprotected springs, rivers and other sources.

Regarding the frequency of travel to fetch water from the source, 48% of the total samples have replied that they made two trips a day. Those who have made a single trip per day constitute 21% of the total respondents. Respondents representing 16% made 3 trips per day while the remaining 12% made more than 3 trips a day to fetch water from the source. About 86% of respondents use both human and domestic animal power to transport water (see **Figure 3**).

Although most of the water supply schemes considered in this study are located at a convenient place and reasonable distance, the average waiting time/queuing is high and the findings indicate that queuing time at the water source is a major issue that takes away the labor force necessary for other productive activities. About 37.8% of the respondents spend up to 30 minutes while 9.5% spend from 31 - 45 minutes as queuing time. More than 49% spend more than 45 minutes as queuing time at the water points.

The average house hold water consumption pattern depends on the family size, the level of income, economic activity, the degree of engagement in productive activity and consumption behavior. The household level economic and productive activity determines the amount of water used per day for the different purposes. The findings of this study indicated that 11% of the respondents use up to 40 liters per day for household cooking, drinking, sanitation and others purposes while 24% consume 41 - 60 liters per day. Similarly, about 24% and 26% consume 61 - 80 liters and 81 - 100 liters per day respectively while 15% consume more than 100 liters of water per day. In Ethiopia, about 20 liters of water per person per day and accessible within a range of 0.5 to 1.0 km from a dwelling place is considered as adequate water supply [22]. Any improvement made in safe water access has to be evaluated as per this definition. Taking this definition in to account, Dessalegn [8] has stated that significant proportion of rural households with safe drinking water access will have less chance of securing adequate amount of water that is quite necessary for their wellbeing. Considering the average house hold number (approximately 5) in the study area, only about 15% may get 20 liters of water per day per person. This indicates that the water supply is inadequate, even though the schemes are accessible at reasonable distances. The findings of this study indicated that some of water points are also used for watering livestock. About 23% of the respondents use ponds and/or water taps to water their livestock.

Table 2. Findings of the focus group discussions sessions conducted at 4 study communities.

Issues of group discussion	Summary of collective responses of 4 focus group discussions participants on issues under discussion					
	0	1	2	3	4	Scoring
The rate of community participation in planning and implementation of the water supply project	No participation	Very limited participation	Participation rate was fair	Participation was good	Rate of participation was very good	4, 4, 3, 4
Community participation on choice of technology used for the water supply scheme	Not at all	Participation was very limited	Participation rate was fair	Participation rate was good	Rate of participation was very good	0, 0, 1, 0
Technical and capacity building training given to water committee members	Not at all	Yes once	Yes twice	Yes three times	More than three times	1, 1, 1, 1
Do water committee have a regular basis of reporting systems to the concerned body	Not at all	Yes, to the community, KPA, district when there is a problem not continuously	Yes, to the district water office	Yes, to the KPA administration	Yes, to the district water office and KPA administration but not to the community	4, 1, 1, 1
Obtaining drinking water from your present source with less effort than the former water source	Not at all	Very limited advantage gained	Yes, relatively speaking	Yes, in many aspect	Yes, in very many aspect	4, 4, 0, 3
Collection and management of water fee, and control mechanism. Community regularly informed on how the money is kept and spent	No collection & management of water fee	Water fee is collected by the vendor but mismanaged by the water committee	Water fee is collected by the vendor but water committee reported to district water office only not to the community	Water fee is collected by the vendor but not reported to district water office & KPA administration	Water fee is collected by the vendor & reported to the district water office & KPA administration but not to the community	4, 2, 0, 4
Benefits of the improved water source	No visible improvement and benefit to the community	Very limited improvement and benefit to the community	Improvement in household health and increase in productivity	Improvement in household income and	Improvement in household health, income, labour productivity, live stock &long distance travel to fetch water	4, 4, 0, 4
What are the major problems of your water supply scheme	There is no significant problem at present	No access to safe drinking water & rely on unsafe sources Hand pumps are not working throughout the year	Limited water taps to address the ever increasing population growth	Pipe line breakage, seasonal fluctuation and decreased flow during dry season	Pipe lines stolen, limited pumping power ever increasing fuel cost, limited water taps for ever-growing population, investors competing for consumption from the same water supply system	4, 0, 1, 3
Suggested solution to improve and sustain the operation of the existing water supply scheme	No solution for the moment	Ensure access to safe drinking water & maintain the existing hand pumps	To change the diesel generator to hydro power, the existing source should be complemented by additional source	Ensure the safety of pipelines extended from adjacent district facilitate maintenance	But we need to have additional water taps to cope up with population growth and income generating activities at household level	2, 4, 1, 3
Follow up and monitoring mechanism of operation and management of the water supply scheme	No monitoring mechanism in place	Very little practice of monitoring and follow up	There is some level of monitoring	There is good monitoring and follow up mechanism	There is very good monitoring and follow up mechanism	0, 0, 1, 1
The role KPA administration concerning the water supply	They should not involve	They should involve in the management of the water supply system	KPA administration has to work closely with the water committee	KPA administration take the lead in the management	Coordinate & facilitate operation and maintenance of water supply	1, 2, 2, 2

KPA—Kebele Peasant Association.

Table 3. Findings of the key informants interview sessions undertaken at 4 study communities.

Issues of interview session	KI Participants collective responses of the interview session					
	0	1	2	3	4	Scoring
Community members make payments timely and it is sufficient to cover operation and maintenance cost of the water supply schemes	No, we have not yet accessed with safe drinking	Yes, but fee is not sufficient for operation &maintenance	Yes, relatively sufficient for operation and maintenance	Yes, fee is sufficient for operation and maintenance	Yes, fee is very sufficient for operation and maintenance	4, 4, 0, 4
Benefits of the improved water source	No benefit as we don't have access to safe drinking water	Very limited improvement and benefit to the community	Improvement in household health and increase in productivity	Improvement in household income and health	Improvement in household health, income, labor productivity, livestock &long distance travel to fetch water	4, 4, 0, 4
Efforts made by NGOs to strengthen Water committee	Not all	Very limited	Satisfactory	Encouraging	Very encouraging	0, 1, 1, 4
Collection and management of water fee, and control mechanism. Community regularly informed on how the money is kept and spent.	No collection & management of water fee as we have not accessed with safe drinking water	Water fee is collected by the vendor and everything is managed by the water committee and community is not informed	Water fee is collected by the vendor and water committee reported financial status to district water office only	Water fee is collected by the vendor but not reported to district water office & KPA administration	Water fee is collected by the vendor & reported to the district water office & KPA administration only	1, 1, 0, 2
Obtaining drinking water from your present source with less effort than the former water source	Not at all	Very limited advantage gained	Yes, relatively speaking	Yes, in many aspect	Yes, in very many aspect	4, 4, 0, 4
Community participation in water supply scheme planning and implementation	No participation	Very limited participation	Participation rate was fair	Participation was good	Rate of participation was very good	4, 4, 3, 4
Responsible body for maintenance of the water supply schemes	No one is assigned	Someone from water committee	Water committee	KPA and Water committee	District water & energy office	1, 2, 0, 4
Women participation in planning and management of water supply scheme	Not at all	Participation is very limited	Participation is satisfactory	Participation rate is good	Participation rate is V. good	2, 1, 1, 2
Major problems of your water supply scheme	There is no significant problem at present	No access to safe drinking water & rely on unsafe sources. Hand pumps are not working throughout the year	Power blackout, pipeline breakage by flooding	Pipe line breakage, seasonal fluctuation and decreased flow during dry season	Pipe lines stolen, limited pumping power, ever increasing fuel cost, limited water taps, water consumption rate of investors	2, 4, 1, 3
Suggest solution to improve and sustain the operation of the existing water supply scheme	No solution for the moment	Ensure access to safe drinking water & maintain the existing hand pumps	To change the diesel generator to hydro power be complemented by additional source	Ensure the safety of pipelines facilitate actual and preventive maintenance, additional tanker and changing PVC pipes with iron pipes	Diesel generator replaced by hydro, additional taps to cope up with population growth, additional tanker, investors need to have their own water supply	3, 4, 1, 3
Follow up and monitoring mechanism of operation and management of the water supply scheme	No monitoring mechanism in place	Very little practice of monitoring and follow up	There is some level of monitoring	There is good monitoring and follow up mechanism	There is very good monitoring and follow up mechanism	1, 1, 0, 3
The role of KPA administration in the operation and management of the water supply scheme	They should not involve	They should involve in the management of the water supply system	KPA administration has to work closely with the water committee	KPA administration take the lead in the management of the water supply scheme	Coordinate & facilitate operation and maintenance of water supply	4, 2, 2, 3

4.3. Community Attitude toward Safe Drinking Water, Sanitation and Hygiene

Respondents' perception towards safe drinking water has got an interesting dimension in terms of understanding their level of awareness concerning hygiene related issues. It is also an indication of planning and implementation of hygiene related activities where issues pertaining to water and sanitation are well addressed [23]. Some of factors affecting the safety of water in the communities under consideration have been identified in this study. These include contamination by livestock, bird and wildlife, flooding, human feces, and algae development (see **Figure 4**). Water sources need to be protected and safe guarded from possible agents of contamination. There are possible health hazards which are associated with exposure of the water supply source to human waste, wild life, livestock and flooding [24]. The safety and quality of drinking water is further in jeopardy as the culture of open defecation has been socially accepted and widely practiced in most of the Ethiopian rural settings and partly in urban areas as well [7].

Figure 3. Donkeys and camels are widely used for transporting water in the study area.

Figure 4. Unprotected and broken hand pump structure in Bubissa Kussaye community.

The study area under consideration is located in Awash River basin which has got complicated water quality problems that calls for special attention to be given by stakeholders involved in various agro-industrial activities. This is due to the fact that Awash River is highly exposed to pollution caused by untreated waste-water discharges from domestic and industrial activities undertaken in Addis Ababa and surrounding areas [25].

As the population in the downstream extensively uses the water from the river for drinking, irrigation and other socio-economic activities, associated public health risks are significant. The occurrence of high fluoride concentration in groundwater in and around the Awash river basin is another public health concern that requires due consideration of development stakeholders in the area [25]. The indiscriminate use of different kinds of pesticides by small and large scale irrigation schemes, and chemical discharges from agro-industrial facilities and smallholders have contributed a lot for surface and ground water contamination of the area under investigation.

Responses given by respondents concerning the water safety issues have indicated community members' level of awareness regarding water safety and what need to be done in the future. It was well noted that maximizing the safe drinking water coverage in rural area is highly associated with the reduction of water born disease and improvement of health status leading to increased household productivity. This can be ensured by increasing awareness of household sanitary situation and working towards improving the sanitary facilities at the household level. Attitudinal change and practicing improved method of waste disposal can greatly improve the safety of drinking water and thereby maximize productivity [24].

The study result has revealed that much has to be done with respect to improving the sanitary facilities at the house hold level. This situation need to be reversed as only 3.4% of the respondents use ventilated and improved pit latrine (VIP). About 45% and 9% use traditional pit latrine and pit latrine with slab respectively. The open defecation accounts for 39% respondents indicating that immediate intervention is required. In Ethiopia, about 12% of the total population use improved toilets, 7% use shared toilets, 21% use traditional toilets and the remaining 60% practice open defecation (8% urban and 71% rural) [26].

4.4. Community Participation in Water Supply Scheme Planning and Construction

Table 4 presents the responses concerning community participation. About 48% of respondents have confirmed that community leaders and members are initiators of their water projects. About 16% pointed out that NGOs, governmental offices and community members made coordinated effort in initiating and developing the water supply project idea while 12% recognized only NGOs and governmental offices as initiators of the their water project.

Table 4. Community participation in terms of initiating water supply project idea.

Responses	Percent
Community members and leaders	43.2
Community members, NGOs & Governmental office	16.21
Government office community members	5.4
Community leaders	4.7
NGOs and governmental offices	11.48
Responsible government office	4.1
NGOs	6.8
Others	6.1
No response	2.0
Total	100.0

Development practitioners working with communities argue that communities need to be given greater opportunity to manage and decide on issues like water supply projects [27]. Community participation enables the beneficiaries to influence the development and management of projects rather than merely receiving a share of project benefits. The shift of emphasis from supply-driven water supply interventions to Demand Responsive Approach (DRA) is due to the fact that the preceding interventions have failed to provide poor communities with sustainable water supplies. In the case of supply-driven interventions, it was found out that beneficiary communities merely take water supply service delivery but failed to play an active role during project implementation and lack a sense of project ownership [28].

Demand responsive approach enable the user community to prioritize their needs, initiate project ideas and entrust informed choices, decide on technology type, and location of facilities that best fit their needs including costs and commitment to bear upcoming responsibilities attached to the project [18].

Community participation in determining the type of technology to be used for the water supply project is found to be about 16%. NGOs and governmental offices are the major stakeholders who decide on the type of technology that need to be installed for the water supply scheme as identified by 60% of the respondents. About 12% of the respondents have assigned the choice of technology to the responsible governmental offices while 5% confirmed the involvement of community members, NGOs & governmental offices. The participation of beneficiary community in choice of technology should be improved. The kind of technology that cannot serve the best interest of the beneficiary in terms of the quality and durability can cause further problem for maintenance and negatively affects the sustainability of the projects [16]. It also could increase the transparency in project development and management.

The community participation goes to the point of contributing cash and labor during the prime time of project construction. Once the water supply scheme is in place, stated the respondents as well as participants of the focus group discussion, "the degree of community participation diminishes living aside all the responsibilities on the shoulder of the water committee members". The water committee took all the responsibilities ranging from collecting water fee to operation and maintenance of the water supply scheme without adequate backup from the KPA administration as well as the beneficiary communities. In most cases, the water committee has no reporting mechanism to the community but they communicate to district water and energy office. The communities involved in this study participated mainly by providing local materials (wood, stone and sand), labor, cash, by active involvement in community consultation, community mobilization and acting as a focal person in those dealings with other stakeholders (see **Table 5**).

The kind of community participations indicated above is the widely practiced form of community participation in Ethiopian context. In general, community willingness to contribute their share of capital costs is crucial to community participation because it acts as an indicator of community commitment to the project [28]. The Community participation could be influenced by combined factors such as the kind of community for whom the project is initiated, the potential benefits to be generated from the project at the household and community level, and identity of project initiators and stakeholders.

4.5. Tariff Collection and Financial Management

According to the information gathered, there is no standard water tariff structure in place in the study area or elsewhere in other rural areas of the Adama district. As per the sample communities, the amount paid for water supplied varies in accordance with the type of technology used for operating the water supply scheme. Those who use boreholes operated by diesel generator pay US$ 0.056 per 50 liters of water while those who use hydropower pay US$ 0.014 per 50 liters. Those using gravity based water supply pay about US$ 0.17 per 50 liters while those using traditional ponds and hand pumps contributes only labor force participating in construction and maintenance.

Table 5. Types of community participation in water supply scheme construction.

Given responses	Percent
Providing labour	7.4
Providing cash & labour	15.5
Providing local materials, labour &cash	29.0
Providing labour and local materials	19.0
Providing local materials (wood, stone and sand)	24.3
Others	4.7
Total	100.0

However, an effective mechanism to administer the water supply tariff in terms of cost sharing practices to cover the ever increasing operation and maintenance cost is crucial for sustained water supply service delivery and benefit over time. This requires a working tariff structure in place that is reasonably framed in order to cope up the ever increasing operation and maintenance cost. The whole issue behind applying appropriate tariff structure is to create a sense of ownership in the community [12] and ensure sustained water supply service over time.

One of the many factors contributing for sustainable rural water supply system is the practice of cost sharing associated with consumers' payment for water service delivered. It was assumed that users should pay as per the water delivered for their house hold consumption. In this respect, the user community is expected to assess the prevailing costs associated with the operation and maintenance of the scheme and thereby revising the existing water service fee as per the ever increasing cost of operation. It is well understood that water users are not in a position to finance the entire replacement costs of their water and sanitation services, it is essential to think about cost sharing as one alternative alongside other new funding mechanism [16].

Table 6 indicates that about 65% of the respondents have practiced cost sharing by paying as per water supplied. There is growing awareness of the respondents concerning the importance of cost sharing paying the water fee on time. They tend to be well aware of the essence of operation and maintenance costs for sustainability of the service. The principle of user pay has been widely practiced in Adama district and the beneficiary communities are willing to pay as per the water supply service delivered. They beneficiaries believe that the fee will be used for covering operation and maintenance cost and building new water scheme. Building new water schemes is required as about 55% of the sample population have stated that the water supplied is not sufficient for the growing population.

4.6. Reliability of the Water Supply Schemes

The issue of functionality of the water supply system and seasonal fluctuations is an area that needs to be given greater emphasis. This is due to the fact that Adama district is geographically located in the rift valley area where the rain fall pattern is sporadic with a corresponding effect of very low ground water recharge rate compared to the amount extracted for use. Drilling borehole and exploiting the ground water potential for water supply has been widely practiced so far. For instance the two water supply schemes (Adullala Hate Aroreti & Geldia Galiye) use boreholes of depth between 150 - 200 meters. The water supply scheme in Adullala community has served for over 10 years and failed to cover the consumption rate of the ever increasing population. Besides, it has got technical problems related to the capacity of the pump and fuel consumption rate of the diesel generator used for pumping the water. The Geldia Galiye scheme is at good status, as the scheme has the capacity to satisfy the consumers demand at its present status. Cheka Dewero scheme used to depend on gravity flown pipe water extended from capped spring located at adjacent high land area in other district. The users of Cheka Dewero scheme have faced the problem of recurrent pipe line breakage and water shortage. That of Bubissa Kussaye (represents 10 manually operated hand pumps existing in 2 communities) depends on traditionally protected ponds complemented by 5 hand pumps installed at different sites. These hand pumps are partly functioning and in short supply of water during the dry season due to low recharge rate of the ground water (see **Table 7**).

Table 6. Response concerning the practice of cost sharing for operation & maintenance of the water supply scheme.

Responses	Percent
Users pay as per water supplied	64.9
Monthly contribution	10.8
Users pay as per water supplied and annual contribution	4.7
Annual contribution	7.4
Others	6.1
No response	6.1
Total	100.0

Table 7. Responses given regarding the functionality of the water supply system.

Given responses	Percent
Well-functioning without any damage or technical problems	11.5
Functioning with some breakage/technical problem	38.5
Partly functioning with some technical problem	35.9
Not functional at all	8.8
No response	3.4
Total	100.0

4.7. Institutional Arrangement

If water projects are to be managed efficiently and are to be sustainable, it is important to promote beneficiary participation in the sense that the main stakeholders should be actively involved in the management of water projects [8]. According to the responses obtained in this study, the responsible bodies to make major decision regarding issues related to the water supply scheme are the water committee (35% of responses), and the beneficiary community (23%). The remaining responses indicated shared responsibility between government, beneficiary community, water committee, and donor agencies. What one can derive from this percentage distribution is that a third of the total respondents used to acknowledge the mandate given to the water committee to manage the water supply scheme in terms of operation & maintenance, collecting water fee, and financial record keeping. The significant participation of the beneficiaries in making major decision indicates that they believe that responsibility should not be left for the water committee alone as such major decisions may have far reaching effect on their livelihoods.

4.8. Women Participation

Increased access to safe drinking water would mean much for women and their children in terms of health, productivity and income. Therefore, the involvement of women has to be maximized in terms of water supply scheme planning, implementation and management [8]. The participation of women in the management of the community water supply schemes under consideration is significant as confirmed by 85% of the respondents. However, the number of women representatives in each water committee is less and not satisfactory. According to the directive given concerning the formation and composition of the water committees in Oromia region, the number of water committee members for water supply scheme is 7 out of which the women representation constitutes only two members [29]. It was widely believed that social, economic and cultural reasons limit the women participation in water committee.

4.9. Consumers' Satisfaction

About 83% of the participants in the study under consideration stated that access to safe drinking water has improved health status of family members. Other benefits recognized by beneficiaries include relief of women and children from exhausting work, more time saved for more productive social and domestic activities, improvement of livestock productivity and household income. **Figure 5** illustrates some major factors that could influence satisfaction of beneficiaries.

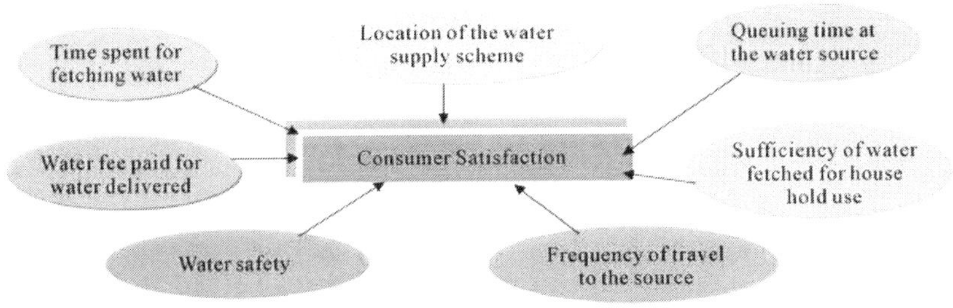

Figure 5. Factors contributing to consumer satisfaction of safe drinking water.

Assessment of data concerning fees has revealed that the water fee is expensive for 26%, cheap for 34%, and fair for 34% of the respondents. About 45% of the respondents stated that they have sufficient improved water supply for drinking, cooking, sanitation and other purposes. However, the majority of the respondents (55%) are still complaining about the sufficiency of improved water supply. Due to the insufficiency, low coverage and seasonal fluctuation of water supply system, the majority of the rural population will have a tendency to use unsafe water sources with high risk of being affected by water born diseases. Unsafe water sources and insecure water storage will create favorable condition for contamination and risk of being exposed to water borne diseases.

4.10. The Role of Beneficiaries to Sustain Rural Water Supply System

There appears greater interest and commitment of the study communities to maintain the water supply schemes and ensure the sustainability of safe water supply services. Active participation of community in project initiation and management could generate favorable conditions for sustainability of the water schemes. About 85% of the total respondents contributed to the water supply system sustainability by actively participating in one or more activities such as consultation and discussions concerning the management of the water supply scheme, maintenance service; cash contribution and paying water service fee in time. Community participation allows the user community to decide on major management issues, costs, investments, and make choices and commit resources as per these choices [18].

4.11. External Support

Once the water committee is formed, the provision of technical training for effective maintenance, operation and management of the scheme need external support in addition to community participation. For this purpose, the involvement of stakeholders that comprise NGOs and GOs is very important [14]. About 45% of the sample respondents have disclosed the importance of communication and collaboration with government offices. In general, about 80% of the respondents mentioned the importance of collaboration with one or more of external bodies such as NGOs, GOs, private sectors, education and research institutions.

The major type of supports given from the district and zonal water and energy offices include maintenance, training, technical assistance represent, provision of spare parts, and financial assistance. The findings indicate that the role of government offices and NGOs focus on filling the gap in areas where the beneficiary community lacks the resources and expertise. The respondents also disclosed that the intervention of external agencies in the management of the water supply scheme needs to be according to the best interest and formal request of the beneficiary community. The beneficiaries intend to have greater say in the management of their respective water supply schemes without being coerced by external agencies [8]. In the study under consideration, some respondents (21%) stressed the necessity of intervention by GOs and NGOs to ensure the sustainability of their water schemes.

4.12. Monitoring and Evaluation

Regarding the management, operation and maintenance of the water supply schemes, the beneficiaries have some sort monitoring system in place. As stated by some respondents, this was a monthly community discussion forum concerning the socio-economic affairs of their respective community where they were given updates regarding the water supply scheme service coverage and operation status. Monitoring at household and community level has to be encouraged as it greatly contributes towards the realization of sustainability. It also creates, for service beneficiaries, an opportunity of being empowered to manipulate their management and performance [16].

5. CONCLUSIONS AND RECOMMENDATION

This study was initiated with objective of investigating the rural water supply systems with case study in Adama area, in central Ethiopia. Both quantitative and qualitative data were collected and analyzed. Four sample water schemes were selected and totally 148 representative households were selected and interviewed. Four focus group discussions were also conducted to gather information necessary for in-depth analysis.

The study assessed issues such as community participation, water committee empowerment, management and governance of water supply schemes, women participation, and functional status of water supply scheme, sanitation and hygiene issues, external support and monitoring system of water supply schemes. The findings indicated that it is important to promote beneficiary participation that gives them greater opportunity to manage and decide on issues affecting their water supply systems.

The study confirmed that the community members have participated well in initiating the project idea, operation, maintenance and management of their water supply schemes, even though their participation in choice of technology required for the scheme was limited. They participated by providing local materials, labor, cash and ideas during consultative discussions. The water committees are established to manage the water schemes in collaboration with concerned government offices and NGOs to enhance the sustainability of the water supply systems.

The water schemes are located at reasonable distances i.e. less than 2 km in most cases and the time taken for round trip to fetch water from source is known to be less than or equal to 30 minutes in most cases. However, the queuing time is more than one hour in most cases indicating that it is a major issue that takes away the labor force necessary for productive activities. Even though the schemes are accessible at reasonable distances, the water supply is inadequate as only about 15% of beneficiaries may get 20 liters of water per day per person.

The cost sharing for operation and maintenance of the water supply scheme is well practiced promoting the financial sustainability of the water supply schemes. However, there are possible health hazards which are associated with exposure of the water supply source to human waste, wild life, livestock and uncontrolled flooding. The indiscriminate use of different kinds of pesticides by small and large scale irrigation schemes, untreated waste water and chemical discharges from agroindustrial firms and smallholders could contribute a lot for surface and ground water contamination of the area under investigation. The issue of ensuring environmental sustainability is out of reach as this requires broader approach of environmental management strategies. Sanitary practices in the study area are poor as only about 3.4% have ventilated and improved pit latrine. The rest use either pit latrine with slab/open pit and/or open field defecation.

The beneficiaries understand well the importance of collaboration with external agencies such as government and non-government bodies, and private firms for effective planning, implementing and managing the water schemes.

Based on the findings of this study some recommendations can be made:

- Providing training for capacity building and refresher training are important in order to scale up the capacity of the water committees to manage the schemes properly. However, the possibility for refresher training is unthinkable in most cases due to the budget constraint at the district level. Therefore, financial and technical support is required not only at community level but at district levels.
- The study revealed that the supplied safe water by these schemes is insufficient. Therefore new schemes as well as expansion of the existing systems should be planned and implemented.
- The study finding has revealed that open defecation is widely practiced in the study area. Therefore, coordinated measures should be taken by concerned government and non-government bodies to provide improved sanitary facilities and appropriate trainings at household level, project level, and community level and district level concerning sanitation and hygiene issues, waste management and protecting water schemes from potential contaminators.
- The present environmental stress in the Awash River basin affects the quality of drinking water in the study area. For this environmental protection measures has to be taken involving governmental agencies, NGOs and the community at large.

6. ACKNOWLEDGEMENTS

The authors would like to express their sincere gratitude to Unicef-Ethiopia for financing this study partially. The community members and experts who participated, in one or other way, during data collection and consultation also deserve due acknowledgement.

REFERENCES

1. UNDP, "Ethiopia: 2010 Millennium Development Goals Report: Trends and Prospects towards Meeting MDG by 2015," UNDP Ethiopia Office, Addis Ababa, 2010.
2. WHO and UNICEF, "Progress on Sanitation and Drinking-Water 2010 Update," 2010. http://whqlibdoc.who.int/publications/2010/9789241563956_eng_full_text.pdf
3. UN World Water Development, "In a Changing World: The United Nations World Water Development Report," 2009. http://www.unesco.org/water/wwap/wwdr/wwdr3/pdf/WDR3_Water_in_a_Changing_ World.pdf
4. ADB, "The African Development Bank in Action: Activities in the Water and Sanitation Sector in Ethiopia; Overview & Key Elements of Interventions," 2010. http://www.afdb.org/fileadmin/uploads/afdb/Documents/Project-and-Operations/2_AfDB_watsan_ETHIOPIA_Au gust_8_Oct.pdf
5. World Water Week, "Presentation from the 2010 World Water Week in Stockholm," 2010. http://www.worldwaterweek.org/documents/WWW_PDF/2010/thursday/T5/2_JMP_latest.pdf
6. WHO and UNICEF, "A Snapshot of Drinking Water and Sanitation in Africa: A Regional Perspective Based on New Data from the WHO/UNICEF Joint Monitoring Programme for Water Supply and Sanitation," 2008. http://www.wssinfo.org/fileadmin/user_upload/resources/ 1251454622-A_Snapshot_of_Drinking_Water_in_Africa_Eng.pdf.
7. D. Aschalew, "Determinants of Household Participation in Water Source Management: Achefer, Amhara Region, Ethiopia," MPS Thesis, 2009. http://soilandwater.bee.cornell.edu/research/international/ docs/Aschalew_THESIS_FOR%20PRINT%20AND%20B OUND.pdf
8. R. Dessalegn, "Water Resource Development in Ethiopia," 1999. http://www.ethiopians.com /Main_FSS_Paper1.htm#rfethio
9. ADF, "Rural Water Supply and Sanitation Program in Ethiopia: Appraisal Report, Infrastructure Department," 2005. http://www.afdb.org/fileadmin/uploads/afdb/Documents/Project-and-Operations/ET-2005-135-EN-ADF-BD-WPREV.1-ETHIOPIA-AR-RWSSP-REV-1

10. EWRMP, "Ethiopian Water Resource Management Policy," The Federal Democratic Republic of Ethiopia Ministry of Water Resources, Addis Ababa, 2001.

11. J. N. Pretty, "Participatory Learning for Sustainable Agriculture," World Development, Vol. 23, No. 8, 1995, pp. 1247-1263.

12. S. Parry-Jones, R. Reed and B. H. Skinner, "Sustainable Hand Pump Projects in Africa: A Literature Review," WEDC, Loughborough University, Leicestershire, 2001.

13. L. Abrams, "Understanding Sustainability of Local Water Services," 1998. http://www.africanwater.org/ sustainability.htm [Paper reference:1]

14. S. Jansz, "A Study into Rural Water Supply Sustainability in Niassa Province, Mozambique," Water Aid, 2011.

15. Water Aid, "Sustainability Framework," 2011. http://www.wateraid.org/documents/plugin_documents/sustainability_framework_final.pdf

16. R. C. Carter, "Study of Sustainability of Water Supply Services in Madagascar," Water Aid, London, 2010.

17. P. A. Harvey, R. A. Reed and B. H. Skinner, "Guidelines for Sustainable Hand Pump Projects in Africa," Interim Report, WEDC, Leicestershire, 2002.

18. J. Sara and T. Katz, "Making Rural Water Supply Sustainable: Report on the Impact of Project Rules," UNDP World Bank Water and Sanitation Program, 1997. http://www.wsp.org/sites/wsp.org/files/publications/global_ruralreport.pdf

19. Oromia Regional State, "Adama District Profile," 2011. http://oromiyaa.com/index.php?option=com_content&view=article&id=250&Itemid=516

20. CSA, "Summary and Statistical Report of the 2007 Population and Housing Census Results of Ethiopia," Central Statistical Agency, Addis Ababa, 2008.

21. N. Mukherjee and C. van Wijk, "Sustainability Planning and Monitoring in Community Water Supply and Sanitation," Water and Sanitation Program, World Bank, IRC International Water and Sanitation Centre, 2003. http://ec.europa.eu/echo/files/evaluation/watsan2005/annex_files/WB/WB1%20-%20Sustainability%20planning%20and%20monitoring.pdf

22. Ministry of Water Resources, "Ethiopian Water Sector Programme," The Federal Democratic Republic of Ethiopia Ministry of Water Resources, Addis Ababa, 2001.

23. WASH, "Financing Water, Sanitation and Hygiene in Ethiopia: Making Every Drop Count Water, Sanitation and Hygiene for All," Tear Fund, 2005. http://www.wsscc.org/sites/default/files/publications/tearfund_wsscc_making_every_drop_count_financing_wash_in_ethiopia_2005.pd

24. NHSSAP, "National Health & Sanitation Strategic Action Plan," The Federal Democratic Republic of Ethiopia, Ministry of Health, Addis Ababa, 2011.

25. G. Taddese, "The Water of the Awash River Basin a Future Challenge to Ethiopia," 2003. http://www.iwmi.cgiar.org/assessment/files/pdf/publications/WorkingPapers/WaterofAwasBasin.pdf

26. AMCOW, "Country Status Overview: Ethiopia," Africa Ministers Council on Water, 2010. http://www.wsp.org/sites/wsp.org/files/publications/CSO-Ethiopia.pdf

27. D. Narayan, "The Contribution of People'S Participation: Evidence from 121 Rural Water Supply Projects," Environmentally Sustainable Development Occasional Paper No.1, World Bank, Washington DC, 1995.

28. E. D. Breslin, "Demand Response Approach in Practice: Why Sustainability Remains Elusive, Discussion Paper," Water Aid Mozambique, 2003. http://www.wateraid.org/documents/plugin_documents/elusivesustainability.pdf

29. OWMERDB, "Amendments of Guidelines for Management of Water Supply Systems," Oromia Water, Mineral and Energy Resources Development Bureau, Addis Ababa, 2000.

2

CHAPTER

Renegotiating the Great Lakes Water Quality Agreement: The Process for a Sustainable Outcome

Gail Krantzberg

McMaster University, Dofasco Centre for Engineering and Public Policy, 1280 Main St. W., Hamilton, Ontario, L8S 4K1, Canada;

ABSTRACT

This is a defining moment for the Great Lakes St Lawrence region, with the opportunity to renovate the regime for ecosystem improvement, protection and sustainability. The binational Great Lakes Water Quality Agreement was first signed in 1972. The outcome of a 2007 review of the Agreement by government and citizens, resulted in a broad call for and revisions to the Agreement, so that it can once again serve as a visionary document driving binational cooperation to address long-standing, new and emerging Great Lakes environmental issues in the 21st century. A prescription for renegotiating the Agreement to generate a revitalized and sustainable future mandates that science inform contemporary public policy, third Party Mediation presses for and coordinates a deliberate negotiation, and inclusive discourse and public engagement be integral through the process.

Keywords: treaty negotiation; Great Lakes; water resource management; public engagement; sustainability

1. INTRODUCTION

Some argue that a collective vision for the future of the Laurentian Great Lakes is embodied in the Great Lakes Water Quality Agreement (GLWQA) signed on April 15, 1972, by Canadian Prime Minister Pierre Trudeau and U.S. President Richard Nixon. The Agreement represents a commitment on behalf of Canada and the United States to restore and maintain the chemical, physical and biological integrity of the waters of the Great Lakes Basin Ecosystem. Article X of the GLWQA Agreement requires the Parties to conduct a "comprehensive review of the operation and effectiveness of this Agreement following every third biennial report of the [International Joint] Commission (IJC)" [1]. The IJC's 12th Biennial Report of 2004, resulted in a science, program and policy review which commenced May 2006 and concluded in October 2007. As of the Spring of 2009, the decision to revise the GLWQA based on the outcome of the review was not publically known. Indications are, however, that the two countries will likely revise the Agreement.

According to Krantzberg [2] this is a defining moment for the Great Lakes, with the opportunity to renovate the regime for Great Lakes St. Lawrence ecosystem improvement and protection. In 2008 we experienced a US presidential election where President Elect Obama advanced promises for restoration of the Great Lakes:

"The U.S. Democrats are proposing a $5-billion trust fund to clean up and restore the Great Lakes if Barack Obama is elected president in November... The $5 billion would pay for a plan that includes designating a Great Lakes co-ordinator to oversee the programs and stepping up the fight against invasive species."

(CBC News, Sept 16 2008)

On November 6, 2008, Rahm Emanuel accepted the position of White House Chief of Staff for Barack Obama. Emanuel is known for introducing in 2003 a far-reaching federal cleanup package under the Great Lakes Restoration Financing Act.

Further, emerging from the 2006 - 2007 review, was a near unanimous consensus that revisions to the GLWQA are not only timely, but arguably imperative. Reporting on the outcome of the Review, the Agreement Review Committee stated: "There was, however, a broad call from the Reviewers for renewal and revisions to the Agreement, so that it can once again serve as a visionary document driving binational cooperation to address long-standing, new and emerging Great Lakes environmental issues in the 21st century." [3].

A significant gap missing in the 1987 GLWQA is the recognition of the requisites for a sustainable Great Lakes St. Lawrence River basin ecosystem. The socioeconomic nature of the region is nowhere acknowledged nor are there programs or policies in place that overtly enhance the economic vitality and social cohesion of the region. The opportunity presents itself to manifest these features into a renewed Agreement that truly integrates human societal needs into the ecosystem approach professed in the current version of the Agreeement.

Should the governments determine that the Agreement be renegotiated; the very process of negotiation will significantly affect the outcome of programs, policies, and participation in Great Lakes revitalization and sustainability. Central to enabling a revised GLWQA to address contemporary threats to the integrity of the region is the rigorous attention to science that can best inform future policy and program direction. Harvey [4] points out that scholars have substantially addressed the role of science in the international environmental treaty-making process [5,6,7]. The Montreal Accord illustrates how science can be central and crucially important in providing policymakers the information to justify action. In setting the basis for an accord to address the depletion of the ozone layer, the international scientific community collaborated with diplomats at the Vienna Convention a few years prior to Montreal, and then at the negotiations in Montreal [8]. This enabled political and economic decision makers to appreciate the science, and to coordinate internationally to assess the need for an urgent and comprehensive international solution to ozone depletion.

Further, a prerequisite to a successful international treaty is a well-informed public that could be mobilized to influence political will. Citizen engagement in the political spectrum can motivate the Parties to negotiate for the benefit of their constituents.

2. THE ROLE OF SCIENCE IN INTER-STATE AGREEMENT NEGOTIATIONS; APPLICATION TO THE GREAT LAKES

Environmental scientists play a key role in society's responses to environmental problems, and many of the studies they perform are intended ultimately to affect policy [9]. Policy makers can make use of science in treaty negotiations to understand and evaluate the nature, scope, scale, impacts and predictability ecosystem phenomena. Harvey [4] describes how the United Nations Convention on the Law of the Sea and the Montreal Protocol, employ scientific data and analysis to facilitate the ability of policy-makers to understand problems with fisheries, climate change, long-range transboundary air pollution, ozone and endangered species. The World Meteorological Organisation and the United Nations Environment Programme acknowledged the importance of science in assessing the problem of climate change when they established the International Panel on Climate Change and held it responsible to evaluate and determine consensus regarding the science behind human mediated climate change.

While policy analysts recognize that policy makers consider political, economic, and social contexts, science is widely recognized as a useful tool for identifying and understanding the problems posed by environmental degradation [6] and thereby informing sound public policy geared towards sustainable futures. For the Great Lakes, a consideration by the negotiating team of contemporary science matters will be paramount to the negotiation of shared priorities to address current pressures that if overcome, would place the Great Lakes on a vector towards improved resilience and sustainability.

Wilson [10] examines three established theories of the science-policy interface important to inform the Great Lakes regime. First, epistemic communities are networks that proceed towards a strong consensus among scientists with clear normative implications and policy relevance that the scientists can rally behind. Generally, though, the majority of environmental problems present themselves with high uncertainty is high so that consensus is difficult to obtain. This is clearly evident in the shared government and academic research on the recent changes to the structure and function of the Lakes Erie ecosystem. Lake Erie is in a stage of transition, reflecting human and environmental influences simultaneously acting on a poorly understood system. Scientists are trying to understand the underlying processes involved in these changes (www.lemn.org, accessed 12/13/08). Once the scientific certainty improves, policy and program priorities can inform the renegotiation of the GLWQA.

The second well establish theory that Wilson [10] presents is Post-Normal Science (PNS), which operates in face of large uncertainty where the policy implications are critical. Within the high risk and high uncertainty context, it is scientific skills rather than scientific knowledge that becomes more important, as knowledge is largely absent. The third theory is Mode Two Science. Wilson explains that in this situation, “science is shifting from a search for truth to a more pragmatic aim of providing a provisional empirical understanding of the world that works in a practical sense.”

In contributing to the science-policy dialogue to inform the negotiation of the GLWQA, scientists must interface with non-scientists to facilitate stakeholder capacity building and involvement as stewards of the natural assets that are the Great Lakes.

There is a complicated feedback relation between science findings and the policy formulation. While maintaining a sort of disinterested objectivity, negotiators of the GLWQA need to include environmental scientists who are cognizant that their research can have policy implication. This nurtures a social responsibility to convey science to lead to more sustainable Great Lakes futures [9]. Since good science is necessary for identifying policy directives, policy makers should seek out scientific research that is conducted professionally, with objectivity, integrity and credibility, and free from political influence and interference [11]. Further, Great Lakes think tanks and expert panels can be commissioned by the governments, to garner results that are not influenced by government positions. Such think tanks can collect evidence regarding specific questions or problems facing Great Lakes ecosystem integrity, make recommendations based on the impartial evidence they gather, and serve an important function for informing negotiations towards a more sustainable international agreement.

3. CITIZEN PARTICIPATION AND CIVIC ENGAGEMENT

The North American Agreement on Environmental Cooperation (NAAEC), the NAFTA environmental side agreement, serves as an important model of citizen participation. The NAAEC established the Commission on Environmental Cooperation (CEC), composed of a Council, a Secretariat and a Joint Public Advisory Committee. The Council is the governing body of government representatives. A trilateral Joint Public Advisory Committee (JPAC) participates in CEC deliberations, including direct interaction with the Council. As Turner [12] describes, the governments each appoint five members to the JPAC, who represent a wide array of stakeholders from industry, academia, and nongovernmental organizations. Each country also maintains governmental and/or non-governmental domestic advisory bodies. Turner [12] elaborates that the NAAEC also contains a public submission process in which citizens may submit claims to the Secretariat regarding the failure of a government to enforce its environmental laws and the Council may direct the development and release of a "factual record" concerning the claim in response.

Inclusivity of this nature is not currently the practice in the Great Lakes Regime. Formed in 1991, in response to the 1987 protocol to the GLWQA, the Binational Executive Committee (BEC) is comprised of senior-level representatives of Canadian and U.S. Federal, State, provincial agencies, Tribes and First Nations. Members are accountable for delivering environmental and natural resource programs and activities that respond to the commitments contained in the GLWQA. BEC is chaired by USEPA and EC and is to meet semiannually as called for under Article X of the GLWQA. Neither the BEC nor the IJC have a comparable Joint Public Advisory Committee (JPAC) or petition process for the Great Lakes region in contrast to the support provided under the CEC.

It is conceivable that in advance of renegotiation of the GLWQA, the governments each appoint equal members to a JPAC, representing industry, academia, and nongovernmental organizations. The inclusion of nongovernment experts in previous GLWQA negotiations sets the precedent for such a JPAC. Learning from other regimes and transferring their experience to the Great Lakes provides a venue for elevating civic engagement in Great Lakes stewardship and sustainability.

A further example of civic engagement in treaty negotiation and implementation is the Columbia River Treaty (CRT) between Canada and the United States of America. The Columbia River Treaty was signed by Canada and the US in 1961 with the purpose of co-operative development of the Columbia River. Part of the public consultation that was carried out with Basin residents resulted in the creation of Columbia Basin Trust (CBT) [13]. In 1995, the Columbia Basin Trust was established with a unique mandate to support the efforts of the people of the Basin to create a legacy of social, economic, and environmental well-being in the region most directly affected by the creation of the Columbia River Treaty dams. The Columbia Basin Trust was endowed with $295 million from the Province of B.C. [14]

One of the priorities for CRT is to inform residents of a renewal or renegotiation of the Treaty when that opportunity occurs (currently set for 2024). The Trust has designated a committee to oversee its programs and appointed an advisory panel of outside experts to assist in the development of its projects and priorities. The Columbia Basin Trust ensures that the views and values of the people of the Basin are a key component of any renewal or re-negotiation process for the CRT.

No such funding entity or citizen vehicle that assures public inclusivity currently exists for the Great Lakes region within or beyond the context of the GLWQA. The precedent, again, is set illustrating that Canada and the United States are open to such engagement infrastructure. The concept of the creation of a Great Lakes Trust is also consistent with the research of Krantzberg *et al.* [15]. The Parties involved in the renegotiation of the GLWQA could build on their experience from other regions of the shared boundary. Consultation among the Department of Foreign Affairs and International Trade (DFAIT), the U.S. Department of State (State Department), across the regions that span the binational boundary could be instructive in the design and process of ensuring public inclusivity.

4. LARGE SCALE ECOSYSTEM COOPERATION

Given the profound changes that will need to be incorporated in a new version of the GLWQA, and given that international water treaties generally take decades to negotiate (the Indus treaty took 10 years of negotiations, the Ganges 30, and the Jordan 40) [16], it is time to launch a basin-wide dialogue to help craft the thinking towards this critical process. The CBT may be a model for serious consideration by the Parties to the Agreement.

Ecological repair of large complex ecosystems continues to be a major challenge. Manning *et al.* [17] find that restoration efforts to date tend to be ad hoc and site or situation specific. The Great Lakes St. Lawrence ecosystem supports rich biological diversity and significant fisheries but population growth and economic development in the Great Lakes Basin during the twentieth century have degraded the environmental quality of the region, threatening its economy and the quality of life of the people who live there. For over 20 years, Canada and the United States have been committed to restoring beneficial uses in Areas of Concern in the Great Lakes [18]. The clean up of the geographic Areas of Concern, represent relatively small-scale efforts that are vitally important to the return of beneficial uses to these locations [19].

However, absent lake-wide or basin wide vision and binational coordination, it is unlikely that this large functioning ecosystem will be achieved through the cumulative effects of small-scale projects. Manning *et al.* [17] point out numerous barriers that prevent large scale ecological restoration projects from being proposed, initiated, or carried through. One barrier is the "shifting baseline syndrome," where the environment experienced by people at one point of time provides the baseline for their expectations regarding quality into the future. If environmental conditions continually degrading, there is an ongoing decline in expectations with each generation as the baseline is lowered. As a result, the expectations of scale and restoration endpoints can diminish over time. Further barriers to the repair of large ecosystems include the scale and complexity of renewal, the long-term and open-ended nature of revitalization, funding impediments, and preemptive constraint of vision [17]. Lack of enforceability compounds the dilemma.

So there are some critical principles requiring consideration for Great Lakes ecosystem revitalization and protection. Complex and lengthy negotiations among numerous government bureaucrats, scientists, politicians and basin stakeholders are requisite for a contemporary and responsive agreement that brings with it a strong likelihood for implementation. The nature of democratic dialogue enables a negotiation that nurtures political leadership based on analytically derived consensus. Such momentum provides a backbone to compile adequate resources for Great Lakes St. Lawrence renewal and protection.

Large-scale ecosystem restoration will not move forward without a comprehensive, science-based plan that includes cost estimates and stakeholder consensus [20]. Further, plans at this level of complexity take a great deal of time and resources. Therefore, purpose, planning, and process definitions are required of the Great Lakes St. Lawrence governments and its citizenry, inclusively defined. To generate the blueprint for the 21st century requires the analysis of complex science, public engagement beyond consultation, arduous negotiation, and a mandatory dedication of sufficient time to get it right.

5. RIPENESS

The present frequency of destructive natural disasters worldwide, is generating heightened public concern for environmental issues, particularly climate change, given the real and potential impacts such stressors and pressures could have on their lives. Harvey [4] contends that this makes it a ripe and opportune time to move on the public's attention and help citizens to push their governments to enact effective environmental policies.

Kingdon [21] contends that the policy window opens when three streams converge. Problem identification and recognition is the first stream, and may be triggered by indicators or specific events. Communities or stakeholders that generate and deliberate on alternatives and proposals is the second stream. The third, the political stream, incorporates modalities in public opinion, different interests as administrations change, and the receptivity of various interest groups. These streams may proceed independently stimulated by distinct priorities. Policy is determined by how government prioritiezes public matters. For public policy to be effectively enacted Kingdon [21] describe the policy primeval soup: where factors align, including technical feasibility, value acceptance, and reasonable change for receptivity. A policy window then opens "because of change in the political stream or... because a new problem captures the attention of governmental officials and those close to them," In essence, a policy window opens in either the political stream or the problem stream leading to coupling efforts on the part of entrepreneurs and a place on the decision agenda.

In the Great Lakes regime, it appears we are at a ripe moment where the political stream and the problem stream are coupling. The 2006-2007 review of the Agreement raised a sense of urgency to revitalize the regime. The recently concluded negotiations on the Great Lakes Sustainable Water Resources Act [22] further heightened attention to the protection of the Great Lakes water resources. In a letter writing campaign, scores of organizations urged the President and the Prime Minister to include Great Lakes priorities in their February 2009 visit, and for the first time in over a decade, the media ceased the cause and wrote of the importance of the GLWQA (e.g. Chicago Tribune, February 12, 2009). Citizens, bureaucrats, and politicians are engaged.

Zartman [23] warns, though, that ripeness is necessary but not sufficient for the initiation of negotiations. It is not self-fulfilling. A ripe time must be grasped directly by the Parties. If not, the persuasion of a mediator may assist progress. Since not all ripe moments are seized and turned into negotiations, Zartman calls on the importance of interested third parties to press for the ripe moment to be ceased, and in this case, to negotiate a renewed GLWQA.

Negotiation literature identifies two strategies for bargaining towards a common agreement: interest-based and cooperative bargaining compared to positional and competitive bargaining. Interest-based bargaining is rooted in the idea that the fundamental interests (or concerns) of the Parties are complementary. For the Great Lakes this would mean that both Parties want to revitalize the ecological, economical and social integrity of the Great Lakes St. Lawrence ecosystem. Interest-based bargaining goes by a host of names such as "consensus bargaining," "problem-solving negotiations," "win-win," "mutual gains," "collaborative bargaining," "principled negotiations," and others [23].

Fisher and Ury [25] define three characteristics of interest-based negotiations relevant to the Great Lakes Regime:

1. Focus is placed on interests, not positions;
2. Options for mutual gain are invented;
3. Objective criteria are used to select the appropriate resolution to an issue.

The identification of interests is also at the center of formal interest-based processes. However, rather than starting from proposals and seeking justification, the interest-based process starts from the justification (the identification of interests) and leads to the joint development of proposals. Once alternative proposals are identified, the Parties, often using objective ranking criteria, select the alternative that ranks highest.

The Conflict Research Consortium at the University of Colorado [26] advises that the Parties, and indeed, any third party mediator should assess the ripeness for negotiating before beginning the process. Questions the two countries should explore are:

1) Are all the Parties interested in negotiating?
2) If not, why is one or more of the Parties reluctant?
3) Can anything be done to make negotiation more attractive to them?
4) Do the Parties all know their alternatives to a negotiated settlement?
5) Do they feel those alternatives are good?
6) Do the Parties have a legitimate leader or representative who could effectively negotiate for them?
7) Is a forum available for negotiations?
8) Is a credible mediator available to assist in the negotiations?
9) Do the Parties agree on the credibility of this mediator?

AFSCME [28] suggest that several factors are needed to determine if an interest-based process is appropriate. Adapting these factors to the Great Lakes St. Lawrence Regime, reveals that the Parties must honestly assess their relationship and the environment in which they operate and ask:

- Does each Party have the authority to negotiate or will the agreements be subject to further review by a higher order of government the executive branch? Legislative review of terminology is likely unavoidable.
- Do the Parties have the ability to clearly and effectively communicate? Good communication is essential to the effectiveness of an interest-based process.
- Is training and facilitation available? Knowledge of the process and third party assistance are necessary for success.

- Are both Parties willing participants in the process? Each party must be motivated to assume the changes and risks associated with interest-based bargaining.
- Does sufficient trust exist between the Parties? The process work best in a mature relationship where the Parties can rely on each others' word.
- Is there internal consensus within the Party to engage in the negotiations? If there is substantial opposition on either side, the process may be undermined. Similarly, if either party's decision makers are experiencing significant challenges to their leadership, the process is not likely to work well.

The Oregon Mediation Center [26] councils for integrative, collaborative or problem-solving approach to negotiation. Spector [29] examines the conditions that facilitate cooperation, often through the mechanism of negotiation. He refers to Druckman's findings that emphasized the importance of situational influences and levers, such as media attention, the number of negotiating parties, firm deadlines, and third party involvement, on facilitating flexibility in the negotiation process and yielding more successful outcomes.

6. PLANNING FOR NEGOTIATIONS

Effective planning is crucial to meeting negotiation objectives. If Canada and the United States are to reach a satisfactory agreement whose provisions will be implemented, specific events that must take place before the Parties "come to the table". These events follow those of Maiese [30], adapted for the Great Lakes St. Lawrence regime.

1. The Parties agree that they have a common problem that they share an interest in solving. This enables the development of a joint delineation of the priorities involved, and the programs, processes and policies needed to resolve them. The way in which Parties define the problem for the Great Lakes St. Lawrence region will fundamentally determine the rest of the planning process.
2. The Parties determine the depth and breadth of their goals, anticipate what they want to achieve spatially and temporally, and prepare for the negotiation process. They define the issues to be discussed and, in consultation with experts and citizens, develop a complete list of the issues at stake.

3. The Parties prepare a combined, comprehensive list of issues and priorities that determines the negotiation agenda.
4. The Parties agree on the location of the negotiations, the time and duration of the sessions, who will be involved in the negotiations, and techniques to pursue if negotiation fails.
5. Lewicki *et al.* [31] advise that the Parties must be aware of their goals and positions, determine which issues are most important, as well as whether the various issues are linked or separate, and be aware of the underlying interests and goals of the other side.
6. The Parties, representing their populations, need to consult with the public and various stakeholders throughout the process. Lewicki *et al.* [31] remind us that negotiators are operating on behalf of a constituency and should consult with their constituents as well as with the other side to ensure that the constituents' needs and priorities are included in the negotiations.
7. Negotiators must be able to present science information and respond to the other Party.
8. Professional negotiators should exchange information about initial proposals and priorities before negotiations begin to inform each Party's intelligence gathering.

7. PARTICIPATORY ELEMENTS FOR SUCCESS

Methods for effectively involving stakeholders in deliberations over science-based policy remain a challenge. Wilson [10] speaks to the importance of the public's own experience-based knowledge, and the need for transparency, involving a broad array of civic perspectives with different interests to gain saliency, legitimacy and credibility. Scholars have advocated the use of participatory techniques to further the involvement of scientists and other technical experts in collaborative environmental decision-making [32]. While the involvement of technical experts is one component of successful collaborative partnerships in the Great Lakes regime, it is not the sole participatory dilemma. The Parties will need to determine how and to what extent to involve the Great Lakes St. Lawrence citizenry in decision-making. The greater the engagement, the greater will be the ownership and commitment towards sustainable solutions.

Day *et al.* [33] find that a process driven solely by technical experts based on contemporary science fails to recognize, access and integrate the diverse values, wisdom, and perceptions of non-technical stakeholders. Scholars illuminate that the involvement of citizens provides a breadth of viewpoints, historical and traditional knowledge, and risk perceptions absent from an exclusive gathering of scientist and technical experts [34,35,36,37].

Scholars have convincingly demonstrated that nontechnical citizens can meaningfully engage in discussions about science and technology [38].

While researchers on public involvement concur on the utility of acquiring views and information provided beyond the scientific community, there is no clear agreement on the most pertinent role of citizens. Nor is it clear how to optimize the design a process of engagement that is most effective in assisting the Parties achieve a shared purpose that is responsive to their constituents input.

Public hearings and meetings are traditional means for the Parties to the GLWQA to solicit feedback on proposed regulatory or policy reform affecting citizens. As a public involvement tool, the value of hearings is limited as dialogue is restricted, often one way, and the platform presents itself as a place to air dissatisfaction leading to a confrontational environment. Trained Mediators should be used, to augment meetings and hearings with interviews with both expert and non-expert Great Lakes stakeholders. Jolley [39] reminds us that non-expert respondents may lack the proper knowledge to respond to questions concerning natural resources policies or problems. Interviewed technical experts and non-technical experts may also have differing perceptions of the risks associated with various environmental problems and how to best address those problems. Once again dialogue, discourse, and debate are lacking, since interviews offer a very weak opportunity for reconciling differences. Table 1reflects the dynamics of different public engagement instruments.

Citizen engagement is different from traditional consultation. True engagement means that, along with traditional decision-makers such as public servants, elected representatives, and influential stakeholders, citizens at large have continuing input into negotiations and greater opportunity to influence outcomes. While traditional consultations tend to provide a snapshot of public opinion, engagement is deliberative, interactive, and on-going. Mendelsohn and McLean [40] state that effective citizen engagement mechanisms permit citizens to be present not only in the beginning and end of the negotiation process, but also in the middle stage where choices and decisions are actually made.

Table 1. Public involvement instruments and their associated characteristics. From [39].

Public Involvement Tools	Degree of Citizen Commitment	Opportunity for Dialogue	Cost
Individual Surveys	Low	None	Moderately Low (dependant on sample size and technique)
Public gatherings for Hearings/Meetings	Moderate	Low (generally uni-directional)	Low
Personalized Stakeholder Interviews	Moderate	Moderate	Moderately Low
Public Deliberation and Engagement	High	High	High

The current the review of the GLWQA can be characterized by being at a stage where government officials listened to the public's view and concerns as provided to them October 2007 by the way of the Agreement Review Committee's reports [3]. These officials provided to feedback to the participants who contributed to the review findings report. With no information forthcoming on whether and how the Agreement will be renegotiation, there is a reasonable probability that Party officials will accept the commentary provided by citizens, but make decisions internally. Such a consultation process, then would be considered "venting", a process where citizens and stakeholders provide views with no feedback. Alternatively, the Parties could convene public meetings to explained or justify the decisions taken, a consultation technique referred to as "telling and selling" [40].

Successful collaboration, however, helps participants recognize that they must work together when participants concentrate on shared goals, common problems, a sense of crisis, or a strong sense of place [43]. This applies clearly to the Great Lakes region, where citizens and stakeholders hold a strong identification with a geographic location. Almost all primers on negotiation argue the need to focus on interests not positions [43].

Mendelsohn and McLean's [40] concept of engagement would well suite the Great Lakes St. Lawrence Region. They describe citizen engagement as an open-ended yet structured public dialogue that contributes to specific decisions in a transparent, publicly accountable manner. To further these goals, they recommend a jointly appointed council responsible for enforcing, monitoring, and reporting on progress related to the citizen engagement provisions.

For the Great Lakes regime one could envision an Agreement renegotiation process that is lead by the two federal governments, advised by a multi-stakeholder panel consisting of a broad cross section of interests spanning industry, environment, academia, business, citizens and the list continues. Such a panel would help share the shared vision, deliberate on priorities, principles, and governance, and hold the parties accountable to their commitment. The jointly appointed council would ensure the parties maintained their commitment to shared decision making. This negotiation oversight council would conceivably be responsible to both orders of government, perhaps directly to a Ministerial (or Administrator) Council, and would make recommendations on what other kinds of engagement activities to use in particular circumstances. Such an approach would be well suited to the Great Lakes St. Lawrence region. The collaborative decision making would also generate shared responsibility for implementation of the programs and support for the policies in a renewed and renovated Agreement.

8. THE GREAT LAKES WATER QUALITY AGREEMENT: THE FUTURE AT STAKE

At the cusp of the 1999 aborted review of the GLWQA, Donahue [41] wrote:

"The Great Lakes Water Quality Agreement is a road map to our destination: a restored and enhanced basin ecosystem that meets the general objectives as presented in Article III. (GLWQA supra) It has been twelve years since we have examined that road map and, while the destination remains the same, we are certainly not where we were in 1987. We can not move forward unless we are first able to look back, assess progress, evaluate performance and apply lessons learned to the balance of our journey."

It is now 22 years since the Parties renewed that road map, and as in the case of Donahue [41] the regime continues to witness a confusing plethora of Great Lakes institutions, with a pronounced paucity of integration and coordination. Funding levels, policy priorities, and societal preferences have changed. New institutions have emerged, others have evolved, and still others have ceased to exist. Existing institutions are becoming irrelevant [15].

In their report "Promises to Keep: Challenges to Meet" [42], a coalition of Great Lakes environmental nongovernmental organizations recommended several overarching principles that should guide the future spirit and overall thrust of the Agreement, urging that a future Agreement 1) be more preventative in its ecosystem approach (i.e., embrace the precautionary principle), 2) more urgently compel ecological recovery, and 3) provide for more accountability for implementation. To reclaim the Agreement's effectiveness, they state, the Parties must support public transparency in the processes and institutions that fostered public involvement in the past.

Finally, the issue is not whether transboundary groups should be involved in shared protection and restoration of the Great Lakes St. Lawrence. Rather, argue Wondolleck and Lewis [43] the question is how their involvement can be managed so that results are informed and accountable to the broader public trust. Such processes must be accountable by making them complementary to traditional decision making, respecting the rights of public comment and influence, and incorporating independent science and appropriate performance measures [43].

Kingdon [21] concurs that in the end, public policy must be realistic and compatible with socially held values; in this context, current public policy on Great Lakes St. Lawrence programs and priorities must more precisely reflect what the Great Lake St. Lawrence regime has said during its review of the Agreement, that being to reflect contemporary threats and opportunities for binational innovation.

From the scholarly literature to practical experience in the Great Lakes St. Lawrence regime, a prescription for a revitalized future for the GLWQA mandates:

- Science information to inform contemporary public policy
- Third Party Mediation that reinforces ripeness of negotiation
- Negotiation planning that generates a logical sequence of actions and deliberations
- Inclusive discourse and public engagement throughout the process to enrich the information base and enable implementation
- Establishment of a jointly appointed authority responsible for enforcing, monitoring, and reporting on progress related to the citizen engagement that is responsible to both orders of government, at a Ministerial (Administrator) level.

ACKNOWLEDGEMENTS

The author thanks members of the Great Lakes Futures Round Table and others who have actively participated in the review of the Agreement for their determination to see the Agreement modernized and become the contemporary vision for international cooperation. The author also thanks Douglas Markoff for his tremendous support to make the Lakes Great.

REFERENCES

1. *Great Lakes Water Quality Agreement. 1987. Canada and the United States,* Available online: http://www.ijc.org/rel/agree/quality.html (accessed May 17, 2008).
2. Krantzberg, G. The Great Lakes, A 35th Year Anniversary: Time to Look Forward. *Electronic Green J.* 2008, 26. Available online: http://egj.lib.uidaho.edu/index.php/egj/article/view/3213 (accessed April, 2009).
3. *Agreement Review Committee 2007. Report to the Great Lakes Binational Executive Committee: Volume 1,* Available online: www.binational.net (accessed April, 2009).
4. Harvey, K. Monitoring Change: Citizen Science and International Environmental Treaty-Making.*International Environmental Negotiation: Ensuring a Sustainable Future* **2006**, *15*, 77–90.
5. Susskind, L. *Environmental Diplomacy: Negotiating More Effective Global Agreements*; Oxford University Press: New York, NY, USA, 1994.
6. Andresen, S.; Underdal, A.; Wettestad, J. *Science and Politics in International Environmental Regimes*; Manchester University Press: Manchester, UK, 2000.
7. Wapner, P. *Environmental Activism and World Civil Politics*; SUNY Press: New York, NY, USA, 1996.
8. Highum, E. Remarks Presented at the Annual Meeting of the International Studies Association, Town & Country Resort and Convention Center, San Diego, CA, USA, March 22, 2006.

9. Kriebel, D.J.; Tickner, P.; Epstein, J.; Lemons, R.; Levins, E.L.; Loechler, M.; Quinn, R.; Rudel, T.; Schettler; Stoto, M. The Precautionary Principle in Environmental Science. *Environ. Healt. Perspectiv.* **2001**, *109*, 871–876.

10. Wilson, D.C. *Clearly Recognized and Suitably Porous: A Review of the Literature on the Science-Policy Interface in International Environmental Management. Deliverable no. 1 for SAFMAMS - Scientific Advice for Fisheries Management at Multiple Scales, EU FP6 project no. 013639. Working paper no 1-2007*; Institute for Fisheries Management & Coastal Community Development: Denmark.

11. Krantzberg, G. Dofasco Centre for Engineering and Public Policy. In Presented at Science - policy futures workshop, McMaster University, Canada, May 18, 2006.

12. Turner, J.F. Presented at Assistant Secretary of State for Oceans and International Environmental and Scientific Affairs Testimony before a Joint Hearing of the Committee on Environment and Public Works and the Senate Committee on Foreign Relations, Washington, DC, USA, July 24, 2002.

13. CBTa (2008). Available online: http://www.cbt.org/about/main.asp?fl=3&pg=treaty (ac cessed April, 2009).

14. CBTb (2008). Available online: http://www.cbt.org/water/texts/ColumbiaRiverTreatyB rochure.pdf(accessed April, 2009).

15. Krantzberg, G.; Manno, J.; de Boer, C. Great Lakes St. Lawrence River Governance: EXPERT Workshop. In Presented at the Expert Workshop on Great Lakes St, Lawrence River Governance Convened, Toronto, Ontario, Canada; 2008.

16. Burchfield, J.; McKinney, M. The Columbia River Treaty. Available online: http://www.umtpri.org/documents/ nrcr_program/Practicum_2008.pdf (accessed March, 2009).

17. Manning, A.D.; Lindenmayer, D.B.; Fischer, J. Stretch Goals and Backcasting: Approaches for Overcoming Barriers to Large-Scale Ecological Restoration. *Restor. Ecol.* **2006**, *14*, 487–492.

18. Thompson, R. Status Report of the Commissioner of the Environment and Sustainable Development to the House of Commons. In Presented at the Commissioner's Perspective-2008; Office of the Auditor General: Ottawa, Canada, 2008.

19. International Joint Commission. Status of Restoration Activities in the Great Lakes Areas of Concern. Available online: http://www.ijc.org/php/publications/html/aoc_rep/english/report/pdfs/chapter1-e.pdf (accessed April, 2003).

20. Vigmostad, K.; Mays, N.; Hance, A.; Cangelosi, A. *Large-Scale Ecosystem Restoration: Lessons for Existing and Emerging Initiatives*; Northeast-Midwest Institute: Washington, DC, USA, 2005.

21. Kingdon, J. *Agendas, Alternatives, and Public Policies*, 2nd ed.; Longman: New York, NY, USA, 1995.

22. CGLG (Council of Great Lakes Governers). *Great Lakes - St. Lawrence River Basin Sustainable Water Resources Agreement*, Available online: http://www.cglg.org (accessed March, 2009).

23. Zartman, I.W. *Ripeness*, Available online: http://www.beyondintractability.org/essay/ripeness/(accessed March, 2009).

24. AFSME (2008). Available online: http://www.afscme.org/publications/9684.cfm (accessed March, 2005).

25. Fisher, R.; Ury, W.L. *Getting to Yes: Negotiating Agreement Without Giving In*, 2nd ed.; Houghton Mifflin Harcourt: Boston, MA, USA, 1991.

26. *Good Timing-Identifying "Ripe" Times for Negotiations*, Available online: http://www.colorado.edu/conflict/peace/treatment/idripe.htm (accessed March, 2009).

27. *Interest-Based Bargaining*, Available online: http://www.afscme.org/publications/9684.cfm (accessed March, 2009).

28. *Competitive and Integrative Approaches*, Available online: http://www.internetmediator.com/medres/pg25.cfm (accessed March, 2009).

29. Spector, B.M. Motivating Water Diplomacy: Finding the Situational Incentives to Negotiate - The Conditions for Negotiation and Cooperation. *Int. Negotiation* **2000**, *5*, 223–236.

30. Maiese, M. *Negotiation*, Available online: http://www.beyondintractability.org/essay/negotiation/ (accessed March, 2009).

31. Lewicki, R.J.; Saunders, D.M.; Minton, J.W. *Negotiation*, 3rd ed.; Stanford University Press: San Francisco, USA, 1999; p. 5.

32. Hinkey, L.M.; Ellenberg, K.T.; Kessler, B. Strategies for engaging scientists in collaborative processes. *J. Ext.* **2005**, *43*, 1–3.

33. Day, J.C.; Gunton, T.I.; Frame, T.M. Towards environmental sustainability in British Columbia: The role of collaborative planning. *Environments* **2003**, *31*, 21–38.

34. Armour, A.M. Risk assessment in environmental policymaking. *Policy Stud. Rev.* **1993**, *12*, 178–196.

35. Beierle, T.J. Using social goals to evaluate public participation in environmental decisions. *Policy Stud. J.* **1999**, *3*, 75–103.

36. Fiorino, D.J. Citizen participation and environmental risk: A survey of institutional mechanisms.*Sci. Technol. Hum. Val.* **1990**, *15*, 226–243.

37. Isaacson, P. Pollution regulation and public sensibility. *Environ. Impact Assess. Rev.* **1986**, *6*, 229–232.

38. Hagendijk, R.; Irwin, A. Public Deliberation and Governance: Engaging With Science and Technology in Contemporary Europe. *Minerva* **2006**, *44*, 167–184.

39. Jolley, G.J. Public Involvement Tools in Environmental Decision-Making: A Primer for Practitioners. *J. Ext.* **2007**, *45*, 2–3.

40. Mendelsohn, M.; McLean, J. 'Values and Interests in Attitudes toward Trade and Globalization: The Continuing Compromise of Embedded Liberalism,'. In *Ruggie*; Gerard, John, Ed.; Ashgate Publishing: Farnham, Surrey, UK, 2008.

41. Donahue, M.J. The case for good government: why a comprehensive review of the Great Lakes Water Quality Agreement is needed. *Toledo Journal of Great Lakes' Law, Science & Policy* **1999**, *2*, 1–11.

42. *Great Lakes Forever*, Available online: http://www.greatlakesforever.org/ (accessed April, 2009).

43. Wondolleck, J.M.; Yaffee, L.S. *Making Collaboration Work: Lessons From Innovation In Natural Resource Management*; Island Press: Washington, DC, USA, 2000

3
CHAPTER

On Prerequisites for the Application of Sustainable Development Indicators in Urban Water Management

Ulrika Palme

Department of Energy and Environment: Environmental Systems Analysis, Chalmers University of Technology, SE-412 96 Göteborg, Sweden

ABSTRACT

Semi-structured interviews with 47 key actors were conducted in Swedish water utilities on why Sustainable Development Indicators (SDIs) are or are not used. Important influencing aspects identified included organizational inertia, social capital, the national water sector and authorities. Divergent views of SD and indicators appear to hinder SDI initiatives. Possible explanations are that: (a) not all actors look at decision-making as the kind of rational process the focus on indicators implies, and (b), Swedish urban water systems are widely regarded as sustainable. The water sector itself and regulation are identified as the strongest potential drivers for increased use of SDIs.

Keywords: sustainable development indicators; urban water systems; decision-making; field studies

1. INTRODUCTION

Over the past 15–20 years, sustainable development (SD) has become a generally acknowledged, though distant and elusive, goal for all levels of society to strive towards. Interim targets, exemplified by the United Nations Millennium Development Goals [1] are often used to make the objectives specific and concrete. Furthermore, to monitor and assess progress towards the targets and help steer development in the desired direction, sustainable development indicators (SDIs) have been developed for use at many different levels: internationally, for example, in the United Nations [2], EEA [3], and OECD [4]; nationally in several countries, including Sweden [5] and the UK [6]; and at the community level. Initiatives at the community level include the example of the Boston Indicator Project [7], in which local individuals are involved in SDI development, as well as initiatives on SDIs to be applied within public administrative bodies [8,9]. Finally, SDIs are also frequently applied within business organizations, where one of the most influential actors, and a useful source of information, is the Global Reporting Initiative [10].

Water supply and sanitation are services fundamental to developed societies and indicators covering water are included in most sets of SDIs. From community to international levels, common examples of such SDIs are the percentage of the population with access to drinking water and sanitation, water extraction (sometimes in relation to available water resources), water reuse, and levels of phosphorus and nitrogen in surface waters. At the organizational level, generally speaking, indicators capturing water issues are primarily those covering water consumption and emissions to water. There are also numerous recommendations regarding sets of SDIs designed specifically for use in water utilities, *i.e.*, for urban water systems [11,12,13,14,15,16,17,18,19,20]. These sets describe the activities of water utilities (water supply and sanitation) in more detail, capturing aspects such as resource use, emissions, ecosystem impact, reliability, robustness, flexibility, cost of operations, investment in innovation, research and development, health and safety, participation, and community development. There are, however, few reports dealing with the practical application of SDIs in water organizations around the world.

This scarcity of reports on the application of SDIs in managing urban water systems prompted the project described here, to investigate, through in-depth field studies, the preconditions for sustainable development information systems (including SDIs) in Swedish water utilities. This paper reports and analyses the results from these field studies, to explore the key factors influencing the application of SDIs in

Swedish water utilities by identifying the drivers for and barriers to their use. The paper adopts a social constructivist approach to research methodology and focuses on describing the use of indicators in an organization from the perspective of the various actors constituting the organization. While this research is relevant primarily to Swedish circumstances, preliminary discussions of the results with an international audience has shown that many of the conclusions have wider validity for planning and decision making in urban water management.

This introductory section proceeds by describing the rationale used in examining sustainable development indicators, and also gives a brief account of the Swedish water sector. Subsequent sections describe the methodology used for the field studies and present and discuss the results.

1.1. Sustainable Development Indicators

Within the overarching rationale of using indicators to guide development away from unsustainable trajectories towards more sustainable ones, SDIs can serve different (sometimes overlapping) functions:

- depicting current conditions, evaluating various management actions for the future, and warning of impending changes [21],
- planning and communication [22,23],
- learning, structuring understanding, and conceptualization [24,25], and
- expanding, correcting, and integrating worldviews [26].

Notably, indicators, generally speaking, are routinely used for planning and control in organizations. The functions of indicators described in the management literature primarily correspond to those in the first two bullet points above, although the vocabulary is different: accounting, reporting, bench-marking and management by objectives. For a review of functions of indicators, including SDIs, see e.g., Palme [27].

An aspect of indicators often disregarded, or at least not explicitly mentioned, is that they are linked to a view of decision-making in organizations as a rational process [28]. Rational decision-making, according to Hatch [28], Thompson and Tuden [29] and Thompson [30], is characterized by agreement within the organization on the goals to be achieved as well as on methods that should be used to reach these goals.

As sustainable development issues are often characterized by high levels of uncertainty and conflicting goals, such agreement is often not at hand, which implies an inherent contradiction in the concept of SDIs. Similarly, the kind of rational thinking implicit in much of the sustainable development literature is based on a linear view of strategy development, emphasizing top-down processes, planning and goal achievement [31]. This, however, is not necessarily how decisions are made in organizations [32]. If, instead, strategy is regarded as adaptive (or emergent), with bottom-up processes complementing the top-down processes, planning is still possible and indicators are still potentially useful; However, the view of strategy as adaptive, constantly responding to unexpected events and a changing external environment [27,30], makes indicators less valuable and reliable. Furthermore, the view of strategy as *interpretative* [31] leaves little room for indicators: the surrounding world is regarded as so unpredictable, and the interpretation of it so subjective, that strategic intent can only be executed through symbolic action. Mintzberg [33] elaborates further on the connection between uncertainty and strategy, arguing that a rational model of strategy is relevant for situations with low complexity and low pace of change. This further highlights the inherent contradiction in the concept of SDIs as this description of reality is not typical for most sustainable development decision-making contexts.

The paradoxical nature of using SDIs in decision-making and strategy supports the idea that promoting learning and structuring understanding is possibly the most important function for SDIs, as argued by e.g., Rydin *et al.* [34] and Innes and Booher [25].

The intended function of an SDI, along with its intended users—providers and receivers of information—will affect what information is carried by the indicator, and to what extent this information is aggregated. As described by Mitchell [35], there are two main approaches to indicator design: (1) indices where information is aggregated into one single variable, e.g., the index of sustainable economic welfare (see Mitchell [34] for more examples), and (2) sets of indicators, including many variables, that in their entirety capture the various dimensions of the SD concept.

The purpose of indices within an organization may be explicit, but single indicators making up a set of SDIs, as described above, may be no more than a conventional performance (or service) indicator when taken out of the set. For example, a common indicator in the water sector is mg/l of nitrogen in the effluent from the wastewater treatment plant. This is normally regarded as an environmental performance indicator rather than an SDI (*i.e.*, an indicator used to monitor, assess and steer progress towards sustainable development), but certainly the emissions to water of nitrogen are highly relevant from a sustainable development perspective. Because of the difficulties at the organizational level of distinguishing SDIs from various (key) performance indicators, it was decided in this work to consider a very broad range of indicators, including all that have been recommended and/or applied in the water organizations studied to convey information concerning any aspect of sustainable development *and*which form part of a set of indicators that in its entirety also captures other aspects of sustainable development. The definition used is deliberately broad so as to include all relevant information. Furthermore, SDIs are not regarded as necessarily novel but include developments of existing indicator sets.

1.2. The Swedish Water Sector

The Swedish water sector is entirely public, and its constituent organizations are either publicly (municipally) owned companies or parts of local public administrations. Altogether, 99% of the sector's costs are covered by tariffs, smaller municipalities sometimes subsidizing service provision out of local taxes [36].

The activities of Swedish water organizations are supervised by a range of authorities, which implies a great deal of mandatory reporting. At the local level, municipal committees for environment and health are responsible for drinking water quality, water protection and, in the case of small plants, permits to discharge treated sewage. The County Administrative Board is responsible for water protection at the regional level, and issues discharge permits for most plants. The central supervisory agency for drinking water quality is the National Food Administration, while the Environmental Protection Agency is responsible for the protection of water resources. Discharge permits for the largest plants are issued by the Regional Environmental Courts [36] .

By and large, the design of Swedish urban water systems follows what Wilderer [37] calls "the classical concept of urban water supply and sanitation," according to which the flow is linear from the source (the water reservoir) to the sink (the receiving water) and the main purpose of wastewater treatment is conversion and destruction of materials rather than recovery. However, the aim of increased nutrient recycling is set forth in the Swedish Environmental Objectives, according to which phosphorus should be recycled from urban to rural areas without risking human health or the environment [38]. This is part of an overarching strategy for achieving resource-efficient material cycles free from hazardous substances.

2. FIELD STUDY METHODOLOGY

The research described here followed what Yin [38] calls a holistic multiple-case design, which means that the study consists of several (three) cases and that each case (field study) is a unit of analysis. Municipalities were selected for the field studies according to two criteria: (1) they should possess water systems representative of Swedish urban water systems, and (2) they should include the most important organizational forms assumed by Swedish water utilities. The logic behind the choice of criteria was primarily that of literal replication, *i.e.*, no major differences were expected between the organizations. It was recognized, however, that the design chosen could yield contrasting results due to differences in basic water system features or organizational form. If the results proved to reveal such differences, additional field studies could be made in future to explore any patterns, *i.e.*, following a logic of theoretical replication where contrasting results are expected for predictable reasons [39].

The methodological perspective is one of social constructivism to the extent that organizations and their use of information are regarded as social constructions. This implies emphasis on qualitative rather than quantitative results [40]. Interviewing was therefore chosen as the central research method in the field studies. Quotations from these interviews constitute the raw data of the study and are used in the presentation of the results to convey the interviewees' view of reality in their own words.

Semi-structured interviews were held at interviewees' workplaces, which included waterworks, sewage-treatment works, and offices. Interviewees in the chosen organizations were selected through conceptually driven sequential sampling or "snowball sampling" [41]. Initially, leading employees from all departments were interviewed; these interviews uncovered matters that needed to be investigated in more detail. Based on these findings, new interviewees were selected who could give more in-depth or complementary information.

In the interviews, the concept of SD was normally approached using existing policies for sustainable development of the organization in question as a point of departure (see interview guideline inAppendix). The subject was then shifted to information used and/or required in that context, and to indicators. Whether or not SDIs were explicitly mentioned depended on how familiar the interviewee appeared to be with the subject. If SDI was not a familiar term, the discussion was focused on economic, environmental and service indicators rather than on SDIs.

The issue of what information should ideally be captured in indicators was approached using an inventory of important flows of information to and from each interviewee, referred to as "network" in the interview guideline in Appendix; "information" is used here in a general sense, not necessarily pertaining specifically to sustainable development. The resulting networks were useful in identifying important actors and information flows, and as a basis for discussing indicator content; they were also useful in identifying factors that were determinant of why certain information flows, SDIs included, were or were not requested, or were even restricted for some reason. Figure 1, Figure 2 and Figure 3show simplified versions of the overall networks built up for each of the organizations as a result of superimposing the individual networks.

The interviews lasted for 30–120 minutes and were normally recorded. Each interview was introduced by asking the informant for consent to (a) record the interview, and (b) publish the results under a chosen level of anonymity and after approval of the Swedish report manuscript [42,43,44]. Notably, a vast majority of informants did not ask for any anonymity, while 1–2 informants in each field study preferred job function and fake initials. The latter was therefore adopted as the level of anonymity for all interviewees. Transcripts of the recorded interviews were sent to the interviewees for additional comments and/or corrections. The returned or, rarely, the original transcripts (if the interviewees did not comment on/correct and return them) were subsequently analyzed by coding all content relevant to the research. In the process of coding, phrases were used as units of analysis, and the codes applied were a mix of descriptive and interpretative codes [41] directly linked to the research questions. The analysis was made in accordance with the open approach applied in grounded theory; *i.e.*, no predetermined classification of answers was used. The key factors presented here were identified mainly through analysis of the answers to the question on SD and information flows in the interview guideline (Appendix, Question 3), but also, as already mentioned, of the discussions around the information networks identified. To enhance the interview results, they were complemented by analysis of strategic documents and observations, an approach described in the qualitative methodology literature as "triangulation" [39,45,46]. All quotations from the interviews and the field study documents, originally in Swedish, have been translated into English by the author.

2.1. The Field Study Organizations

The municipalities studied were Uppsala, Borås and Stockholm. The Uppsala field study was based on 19 interviews and the Borås study on

16; the interviewees in both studies included public officers, works engineers at the waterworks and wastewater treatment plants and local politicians serving on the Sanitation and Streets Boards (in Uppsala and Borås, respectively). The Stockholm Water Company (SWC) study was based on 12 interviews with SWC employees having known experience with and/or responsibility for indicator work in the company, with two politicians serving on the board and one representative of the owner, Stockholm Stadshus AB.

The first field study, in Uppsala, was conducted from April to July 2004. Uppsala is a university town in eastern-central Sweden, close to Stockholm, with a publicly owned and governed water organization serving approximately 190,000 people. Since 1 January 2003, the city's water supply and sanitation have been handled by two organizations—the Sanitation Office and parts of Direct Services—which formerly comprised one organization. The Sanitation Office (a "client-side office") is now responsible for all water and sanitation activities, while Direct Services (a "production administration") has been commissioned to carry out the actual operations. The entire Uppsala public sector is organized correspondingly: the client-side offices and production administrations implement decisions made by local politicians serving on the respective client-side and production boards, in this case the Sanitation Board and the Direct Services Board.

The second field study was conducted from March to May 2005 in Borås, a smaller city with a textile industry tradition in southwest Sweden. In Borås, the publicly owned and governed water sector serves approximately 86,000 people. In contrast to Uppsala, the sector is organized according to a more common Swedish model, in which both water and sanitation services are duties of the Streets Office, overseen by local politicians serving on the Streets Board.

The third field study was conducted at the Stockholm Water Company (SWC) from late October 2005 to March 2006. SWC is a publicly owned company operating the municipal drinking water and wastewater systems in Stockholm and Huddinge. Its activities encompass the production and distribution of drinking water for over one million people and the handling and treatment of wastewater from 900,000 people.

2.2. Present Use of Indicators

Like all Swedish water utilities, the organisations studied also report to authorities regarding the quality and quantity of water, wastewater, and sludge. Furthermore, their activities are subject to national environmental quality objectives that apply at the municipal level, primarily concerning good-quality groundwater, flourishing lakes and rivers, zero eutrophication and a non-toxic environment.

Swedish water utilities also report yearly to the Swedish Water and Wastewater Association (SWWA) on taxes and operations. As an example, a total of 468 parameters (referred to as key ratios) were delivered to SWWA from Borås in 2002–2003 (field study material). Of these parameters, 316 were financial, while the remaining 152 captured aspects of administration, environment, water balance, personnel, operations, quality, and renewal and renovation.

In Uppsala, nine environmental indicators regarding water issues were reported in the municipal environmental report [47], of which, six referred to the quality of the sewage sludge produced, two to the river water quality (nitrogen and phosphorus) and one to per capita water consumption.

In Borås, two initiatives regarding targets and indicators have been launched: one in the local Agenda 21 office (part of the Planning Office and the coordinator of municipal sustainability initiatives) and the other in the Water Division. Though these initiatives are partly linked, they were launched independently. The Agenda 21 initiative is linked to the national environmental quality objectives [48]. At the time of the field study, the initiative consisted of eight targets but included only four indicators relating to water issues. These indicators were very similar to these reported in Uppsala, with two referring to the quality of sewage sludge and two to river water quality. The Borås Water Division initiative, which involves management by objectives, does not explicitly relate to sustainable development; nevertheless, the targets defined largely concern sustainable development, whether implicitly or explicitly [49]. There were also a number of preliminary indicators for following up these objectives. Furthermore, at the time of the study, the head of the Water Division was preparing for the initialization of an environmental management system (EMS), which implies increased application of performance indicators.

Of the three utilities studied, the Stockholm Water Company had by far the most advanced indicator system. The practice of applying standard performance indicators at the company dates back more than 10 years, as is demonstrated by the regular publication of indicator reports [50,51]. Over the last few years, Stockholm Water's indicator system has become integrated with the "integrated management system" (ILS) [52] applied by the city of Stockholm to monitor the five overall objectives established by the municipal council. These overall objectives are to: (1) improve welfare and municipal activities; (2) build housing and develop

Stockholm; (3) make Stockholm an ecologically sustainable metropolis; (4) reverse segregation and deepen democracy and (5) take responsibility for the economy [53]. The Stockholm Water indicator report for 2005 [54] contained 27 indicators for the entire company, presented under the following headings (number of indicators in parentheses): economic platform (12), personnel (8) and energy (5). The remaining indicators are organized into groups pertaining to the various processes in the company, from drinking water production to wastewater treatment. In these groups, the indicators in the report are not further classified, but here they have been roughly categorized to give an idea of what aspects the entire set of process-related indicators captures: operations (17), service quality (13), environment (12), and economic factors (7). Stockholm Water has also developed a customer satisfaction index, built up by the simple addition of two environmental and eight service indicators on a yearly basis [55]. Finally, Stockholm Water has had an EMS since 2003 into which the performance indicators, developed prior to the EMS, have been integrated.

3. KEY FACTORS INFLUENCING WHETHER AND HOW SDIS ARE APPLIED

This section presents factors identified in the field studies as critical for whether and how SDIs are applied in Swedish water utilities; *i.e.*, the drivers for and barriers to using SDIs, as established from the interviews through the analytical process of coding described above. As the entire study has approached the subject of SDIs from *within* the organization, this is also how the presentation is structured; beginning at the core of the organization, and subsequently following a continuum from strictly internal to increasingly external. The quotations selected either represent a frequently expressed opinion, *i.e.*, a widely held view, or the opposite, *i.e.*, an extreme point of view that stands out from more common opinions; which case each quotation represents will be clear from the context.

3.1. Attitude towards SDIs

The (potential) value generally attributed to the use of SDIs (defined as described in the final paragraph of Section 1.1) in the individual organizations can be summarized as low to moderate in Uppsala and moderate to high in Borås and Stockholm. It was obvious in all three organizations that, as will be described below, attitudes varied between the various groups of people interviewed, *i.e.*, operational personnel, planners, and local politicians; naturally, the variation within each of these groups was also considerable too.

3.1.1 Professional personnel

The professional personnel interviewed formed a varied group including works engineers, laboratory personnel, project engineers, and managing officers. All referred to the use of performance indicators as something necessary and well established in the sector for monitoring and controlling operations. The indicators referred to in this context are those capturing primarily technical/operational and economic aspects, including customer service and mandatory environmental aspects. Views on broadening the scope of these indicators to encompass sustainability issues more explicitly covered the full range from "waste of time" to "very good idea". The reason given by those rejecting explicit SDIs was never that sustainability issues were not important; rather, sustainability issues were considered to be so integrated into the activities of the organization that no explicit indicators were deemed to be needed. Goals or targets were regarded as potentially useful to make sure that no important issues were neglected, but SDIs were not. This view was especially widespread in Uppsala, but also appeared occasionally in the other two organizations.

The motivations *for* using explicit SDIs, on the other hand, were primarily their usefulness in supporting development towards increased sustainability (or away from unsustainability), especially as tools for benchmarking and control in management. The latter application was often considered to be neglected.

3.1.2. Planners

People involved in planning, either in the Division of Water and Sanitation (or corresponding unit) or at the level of the local political administration, all expressed a positive attitude towards the use of SDIs in the water sector (or elsewhere, for that matter). The planners often work closely with the decision-makers, the local politicians, who are considered by the planners to be a group of people who think less positively of indicators, as the quotation below from an interviewed city planner implies:

But indicators ... Politicians don't have much faith in them. They just don't. It's more the Streets Office and the Environmental Protection Office that monitor the development with their indicators...

3.1.3. Politicians

Three of the six politicians interviewed confirmed the picture of them as rather negative towards SDIs, as did the representative of Stockholm Stadshus AB (not a politician himself, but employed to implement politicians' decisions). Two politicians were not negative, but stated clearly that they would not themselves initiate an increased use of indicators. Only one of the politicians was entirely positive towards the use of indicators and this person also described himself as somewhat unusual in this respect as compared to his colleagues.

However, the interviewees who expressed a negative attitude towards SDIs emphasized that this was not immutable; rather, their reservations concerned the number of SDIs and how they were designed:

PA: Another thing I find important, is not to drown information in more information. Having a large number of indicators is not automatically a good thing. [...] What's important is that they be good indicators. The organization tends to create a few extra, just to be on the safe side. If nothing else, you'll have a line of retreat, because at least you've done the reporting. But as a politician, if you have a couple of pages of indicators for a department, it takes several years before you figure out how to read them.

PB: For the last ten years now there have been requirements in terms of some sort of quality assurance and reporting with goals and indicators, as a result of which all administrations and companies have worked with them in some way. SWC and some others, are advanced in this aspect ... But I think they may have taken it a bit too far. Right now, we're looking at SWC in terms of efficiency and organization, and, for example, there are so many goals that they can hardly get them all out into the line organization ... They need to scale it back.

The two individuals quoted above were both concerned with receiving too much information, and also, in the case of PB, with the personnel at SWC wasting time on processing indicators. What is considered too much by one person, however, may be perfectly satisfactory for another. The following quotation is from an interview with an SWC board politician (PC), and refers to the same set of indicators that was criticized in the quotation above:

I (Interviewer): The indicators in use, are they good?

PC: Yes, they are very good. They provide indications of development over the years in terms of economy, and also the environment. They are based on goals that have been set, and they are easy to read.

I: So it's a good format?

PC: Yes, very clear and informative.

I: And what about the number of indicators?

PC: Well, you can't take in too much information, but this is good.

There are of course many potential explanations to why the two persons quoted above, PB and PC, have such contrasting views on the indicators in question. An aspect worth considering, however, bearing in mind the comment by PA above that "it takes several years before you figure out how to read them [the indicators]" may be the fact that PC has been on the board of SWC for considerably longer than PB has had his position.

3.2. Time

In all three field studies, the need for practicality was underlined by nearly all interviewees. People at all levels in the organizations studied expressed a fear that more effort spent on SDIs would mean increased administrative burdens squeezed into already tight work schedules. One person chose to turn the question about room for indicators around by questioning other activities in the organization.

There is room, or we can make room [for indicators]. We are already working with goals and indicators. I think that we have to set aside time for what is important. Following up goals is important. We also need to examine our work responsibilities so that we can eliminate less important ones.

This quotation points to a difficulty linked to the use of lack of time as an argument against any innovation: it is in most instances a matter of prioritization. "We have not got the time" or "We cannot afford to" may just be an indication that the informant or their supervisor does not consider a change sufficiently important to justify the time or money required for implementation.

3.3. Environmental Management Systems

In the three field studies, how far the environmental management system (EMS) had developed in an organization was found to be correlated with the organization's general attitude towards indicators. The key features of EMS include goals, and indicators to gauge progress towards these goals (see, e.g., Sheldon [56]). An established EMS could hence be expected to influence attitudes towards the application of indicators generally. The causality could also be the converse: in an organization where management by objectives and indicators to gauge these objectives are established, the adoption of an EMS comes naturally as it follows the same line of logic.

Uppsala was the one of the three organizations where people, on the whole, were least interested in SDIs; this was also the organization where no environmental management system had been applied:

We don't have an environmental management system—we haven't worked on those issues. But we generate indicators every year, and they concern finances, operations, emissions, *etc.* There are many different ones, and of course they have some value... in maintaining a high quality. Indicators can also be helpful when you look at what reinvestments to make.

In Borås, SDIs were ascribed a greater value, and in addition to the two indicator initiatives that had already been started at the time of the field study, preparations were being made for the development of a quality and environmental management system:

My aim is that we should develop a quality and environmental management system for the water and sanitation department. [...] The Public Cleansing division is well on the way in this. They have a full-time employee who can devote all his time to this issue. And I think that's precisely what is needed. During the year I hope to hire someone for that position as well. That will give us a chance to pick up the thread that we have lost in our administration.

Again, the importance of prioritizing the issue by devoting time (personnel) to it is emphasized. It is worth noting in this last quotation that preparations for the quality and environmental management system in Borås are being initiated by a division in the streets department, in a sort of "middle–up and down" process. Finally, in Stockholm, both indicator use and an environmental and quality management system were well established and integrated into everyday practice.

3.4. Organizational Structure

Inertia associated with administrative systems has previously been identified as possibly the most important barrier to advancing sustainable urban water management [57]. In this study, inertia in the flow of information through the organization was identified as problematic in both Stockholm and Borås. In Stockholm, this was mentioned in the context of management by objectives, which does not function if the objectives in question are not known and accepted throughout the entire organization. The interviewee is an employee with responsibilities within research and development projects at Stockholm Water.

I guess it's good to use management by objectives all the way from the ownership to the foot soldiers, so that both Stadshuset AB and the company have some sort of shared description of reality, and an idea of where we are heading. The problem is probably not that we have had too much management by objectives, but rather too little. It got stuck somewhere in the middle, in the management group. They have been very aware of this, they and perhaps the supervisors immediately under them, but it feels like management by objectives has not extended upwards, or properly downwards either.

A somewhat similar situation, *i.e.*, objectives that have little impact because they have become "stuck" somewhere in the organization, emerged in the case of Borås. In Borås, the sustainability objectives as approved by the municipal council, and the lack of monitoring of these objectives, was criticized by one of the employees at the water treatment plant:

I don't think that I have set it [the ambition for sustainable development] higher than the municipal management has, centrally. But an organization like this one has such enormous inertia that, even if I didn't care at all, no one would raise an eyebrow. You can't reach down into the organization. There is no control function when it comes to sustainable development. [...] It's entirely up to the individual whether you want to care about it or not.

Monitoring and following up the *internal* targets of the water and sanitation division appears to function much better, possibly because of the participatory process used in their development and/or because of the proximity to the division head. The following quote is from an interview with one of the officers at the water and sanitation office, but similar statements appeared in most interviews with informants working either at the water or wastewater treatment plant or in the office.

I think it's a bit dangerous to say that it [the water and sanitation department's goal document] is finished. Things can change. I believe the situation is quite dynamic. And you have to be able to run a little check at year-end: have we lived up to these goals? We need to think about these things now and then throughout the year. N [Head of Division] will certainly ask...

Another reason why the *internal* objectives in Borås appeared comparatively functional may be that they did not have to be communicated across any sharp organizational boundary (Figure 2). This stands in marked contrast to the communication of objectives from the municipal council to the professional personnel, or, in the case of Uppsala, between clearly distinct organizations. In the new organization in Uppsala, where what was formerly one department had been split into two—a principal and a "doer" (the two circles in Figure 1)—dysfunctional information flows were associated with the boundary between the two organizations:

If something goes wrong, it's regulated in the order, how we are to report to our customers [the principal]. But in the other direction, when problems are reported to our customers, there are often deficiencies in the information flow. Those routines are not at all as good as when we find out about something. ... What our people [also] lack is information about plans from the Sanitation Office. [...] We often find out that we are supposed to build something only by reading about it later in the newspaper.

3.5. Trust and Conflict

Whether dysfunctional information flows are the cause or effect of mistrust is difficult to tell, but the two coincide. For example, lack of trust or confidence (see Earle and Siegrist [58] for a distinction between the two concepts) extending across the boundary between different parts of the organization (or between organizations, depending on how "organization" is defined) was reported in Stockholm, specifically concerning the relationship between SWC and its owner, Stockholm Stadshus AB (Figure 3). In this example, lack of trust was accompanied by an impression of inadequate communication:

In the management group we have a board, and then we have the owners. And I have to say that communication with the owners has not been all that good. We have not had the exchange of information that we would have needed, and we find that there is a lack of confidence in our organization. We think that we are doing a good job, and that the owners should be able to see that. But the owners do not appear to see it, or at least they don't say anything about it.

Institutional trust is part of the "social capital" that is frequently referred to as a prerequisite for sustainable development in organizations [59,60]. In Borås, one politician explained that he had learnt from the public officers about the values and potential usefulness of indicators, while the other politician interviewed underlined the importance of trust and confidence in the professional employees:

Concerning water and sanitation issues, we have great trust and confidence in the professional personnel. That's necessary as we're dealing with technical issues. After all, I'm a history teacher, I don't know anything about this. I may sound like a water and sanitation engineer to someone who doesn't know these things. But to be honest, I don't really know much about it.

In this case, where the politicians for various reasons are skeptical over the (increased) use of SDIs, the level of trust and confidence they hold in the proponents of SDIs appears to be decisive. The issue is two-tiered as it concerns the idea of indicator use as a tool and/or a process on the one hand, and the conception of sustainable development on the other. In Uppsala the situation was unproblematic as the interest in SDIs and the concern for sustainable development were both comparatively low and evenly shared throughout the organization. In Borås, the views on indicators as well as on sustainable development were more diverse, *i.e.*, the level of conflict was higher (or at least more open), but combined with the atmosphere of trust established in the organization, this conflict appeared to promote action towards sustainable development rather than to stall it. In Stockholm, on the other hand, there was obvious conflict between the owner and (some of) the employees regarding the level of sustainable development to aspire to and how to approach this aspiration, and this tension appeared to be most strongly manifest over the organization's indicator program. Combined with what some of the professional personnel perceived as a lack of confidence from the owners, these conflicting opinions had led to a stalemate in Stockholm at the time of the study.

Well aware that the sample is limited, it is worth noting that the situation in the three organizations described above happens to match the idea in modern organization theory of a correlation between performance and the level of conflict in an organization. This model does not explicitly mention trust but does emphasize related aspects such as levels of cooperation and integration, which are highest at the intermediate, "optimal" level of conflict [28], a situation of which Borås could serve as an example. Both too little conflict, as in Uppsala, or too much, as in Stockholm, lead to lower levels of performance (in terms of progress of sustainable development initiatives, including SDIs, in this case).

3.6. The Municipal Context

The Swedish water sector belongs to the public domain, the individual organizations being either publicly owned companies, like Stockholm Water Company, or part of the public administration, like the water organizations in Uppsala and Borås. Judging from the international literature (referred to in the introduction of this paper), the application of SDIs at the organizational level has advanced further in the private sector. Despite this, the fact that water utilities are municipally owned non-profit organizations was put forward as an argument *for* the application of SDIs by one of the members of the management group at SWC:

A public monopoly must, in my opinion, take every opportunity to present and report on the quality of its operations. Because we cannot measure ourselves on the so-called market in terms of red and black numbers—which is, after all, the method for all other enterprises—we have to express the usefulness, quality, or added value generated by what we do in some other way. As a result, all such measures are incredibly valuable and important for us.

The municipal directives were demonstrated to have a strong effect on the prioritization of sustainability issues by the professional personnel. Several interviewees in all three field studies referred to the official municipal policy documents in these terms: "This is our mission; this is what we have to do" ("commission" in Figure 1 and Figure 2 and "owners' directives" in Figure 3). Notably, the missions referred to are products of the municipal decision-making process. Important actors in that process are the civil servants in the administrative office and the local politicians, including those specially assigned responsibility for water and sanitation. The extent to which the professional personnel from the water and sanitation division (or equivalent) are consulted or included in the process varies from one municipality to another.

The increased use of indicators and management by objectives in municipal activities is part of a transition towards more business-like management of the public sector, a movement known as "new public management" [61,62,63]. According to an interviewee heading one of the water treatment plants, the effects of this transition include an increased administrative burden (similarly, many interviewees expressed a fear of an increased administrative burden associated with SDIs, as was earlier discussed concerning availability of the necessary resources):

The movement of municipal operations towards more "private" performance, but without the resources of the private sector, pushes people into an overly pressured situation.

Several interviewees commented on the increasing administrative burdens they were experiencing, but no one other than the person quoted above made the connection to the "businessification" of the public sector.

3.7. Customers/Users

In the field studies, the customers, *i.e.*, private citizens and enterprises (as regards wastewater treatment, predominantly small and medium-sized enterprises connected to the municipal plant), were often regarded as having very little interest in *how* the urban water system works, as long as it does work from their perspective, *i.e.*, as long as good drinking water is delivered, and drainage and sanitation function. A growing environmental interest among the general populace is, however, expected to make users increasingly interested also in the environmental impact of their urban water systems. The customers were often cited as an indirect driver of SDI application, the direct driver being the importance attached to user information by the professional personnel at different levels of the water organization.

3.8. The Water Sector

The organizations studied were all part of the Swedish water sector, several features of which were revealed in the interviews as influencing whether and how individual organizations apply SDIs.

One important feature of the water sector in this context is that its actors are well accustomed to applying performance indicators to monitor and control operations, as expressed by one of the works engineers:

We follow up extensively on the indicators, but it's something that we've always done—number of leaks per km of pipe, number of stoppages, *etc.*

The fact that water sector actors are well accustomed to using indicators does not appear to be an obvious driver of developing established indicator sets so that they better capture the entire concept of sustainable development. Familiarity with indicators would nevertheless be expected to facilitate their increased and extended use as it means that systems for collecting basic data and for reporting are established and, possibly more important, people are used to the line of logic underlying the use of indicators.

The water sector also consists of its individual constituent organizations. All of the interviewees mentioned other water organizations, or specific individuals in these organizations, as important actors in their networks. Other water organizations are a potential driver of SDI application, especially in benchmarking, although as yet none of the organizations studied specifically benchmark SD issues. The potential of benchmarking projects is however shown e.g., by the participation of Stockholm Water Company in the Scandinavian Six-Cities Group for the performance benchmarking of water and wastewater systems [64]. As mentioned in the section on present indicator use in the organizations, a kind of benchmarking is also carried out on basis of the statistics collected yearly by the Swedish Water and Wastewater Association (SWWA). This data collection initiative focuses on information regarding taxes and operations, though occasionally special enquiries are performed. For 2005, for example, SWWA collected information on a voluntary basis from water and wastewater utilities regarding their use of energy [65]. This project, "The water utilities' contribution to increased energy efficiency in Sweden" (VA-verkens bidrag till Sveriges energieffektivisering), was financed by a grant from the Swedish Energy Agency. The ordinary parameters reported to SWWA, however, are predominantly financial.

An interviewee with long experience of indicator work in the Swedish water sector gave the following account of the background to the collection of statistics by SWWA:

In the 1980s we were attacked by the private sector, by the contractors, who said that they could maintain our systems much more economically than we could ourselves. At that time we didn't even know what it cost to maintain the system. That was actually how it all began. And then we produced a whole lot of financial metrics, but with time we eventually got involved in disturbance indicators, indicators with links to environmental factors and the like, which are actually the ones we use to manage our operations. Especially if you develop a customer perspective. The customer can consist of individual people, but it can also be the environment, so to speak.

The interviewee quoted above describes how the private sector was leading in the measurement of financial performance in the 1980s. The influence of the private sector in this respect could be regarded an impulse leading away from rather than towards sustainable development, considering the one-sided focus on financial performance. On the other hand, the system developed for data collection, reporting, and benchmarking may, as noted above, prove useful in supporting the sustainable development of the sector.

There are many formal contacts (e.g.,, meetings and committees) between water organizations in Sweden, but most of the interviewees especially praised the *informal* contacts. Other water utilities and municipalities were referred to as somewhere to turn for advice, mainly on technical and operational issues but occasionally—and increasingly so—also on the use of indicators.

3.9. National and International Objectives and Regulations

In Uppsala and Borås, the national environmental objectives were obvious drivers of the formulation and application of local sustainable development goals and indicators at the Environment and Health office (Uppsala) and the municipal Environment and Planning office (Borås). In neither Uppsala nor Borås, however, had the water utilities been significantly involved in the respective efforts. Both organizations had reported only parameters that they would have monitored anyway, and the national objectives did not appear to have had any recent and/or obvious effect on either of the water organizations with regard to their use of indicators.

National *regulation* is on the other hand the obvious driver of many performance indicators in use today, as these indicators have to be reported to permitting authorities such as the Environment and Health office (or corresponding unit). This implies that to the extent that national regulation reflects the national environmental objectives, the latter affect the water organizations and their use of indicators via regulation rather than via local SDI initiatives. Despite the obvious influence of authorities on the water sector reporting practices, or possibly because the reporting practices in question are so routine, authorities as well as regulations and objectives at a regional or higher level were mentioned very rarely in the interviews. One of the very few such explicit references was made by one of the politicians in Borås:

The problem is that we don't always know what the new requirements are, like the EU ones. [...] With regard to both nitrogen and phosphorus pollution, which are the most important types, we have to strive constantly to get below the limit values. We don't have planned activities, we're just putting out fires as they arise.

Several interviewees cited the sometimes unpredictable behavior of authorities to explain why the formulation of internal targets may appear futile: an authority's sudden change in the legal or economic premises of operations may render fruitless the effort expended to formulate indicators as well as targets. The broader implications of unforeseeable changes in the water sector are discussed further by Dominguez and Gujer [66].

3.10. Results in Summary

In Figure 1, Figure 2 and Figure 3, some of the factors identified as key in the previous sections have been depicted as drivers or barriers. Factors such as attitudes towards indicators, resources, EMS, and social capital have not been depicted individually, but are reflected by the fact that the organizational core in Borås and SWC functions as an SDI driver, whereas in Uppsala no such driver function could be discerned.

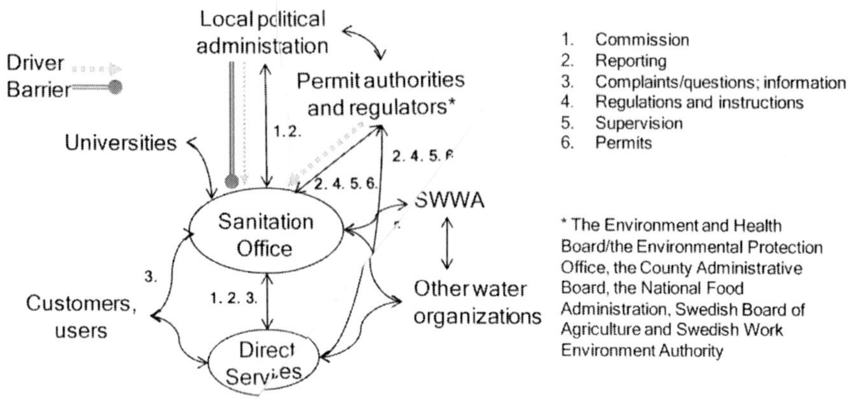

Figure 1. Key actors and information flows in Uppsala water organization.

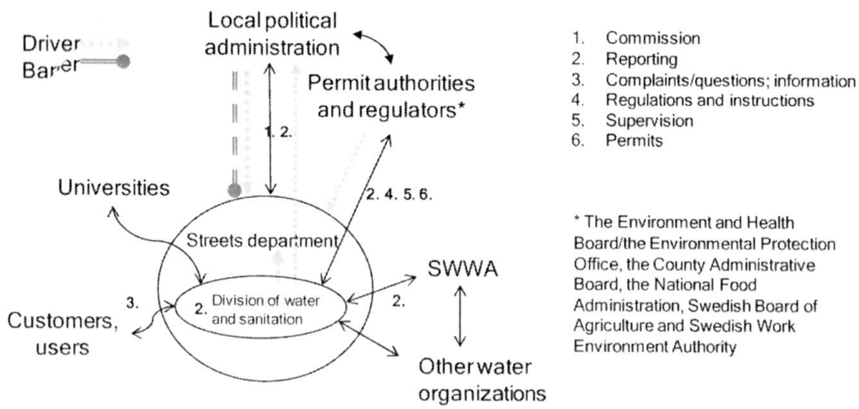

Figure 2. Key actors and information flows in Borås water organization.

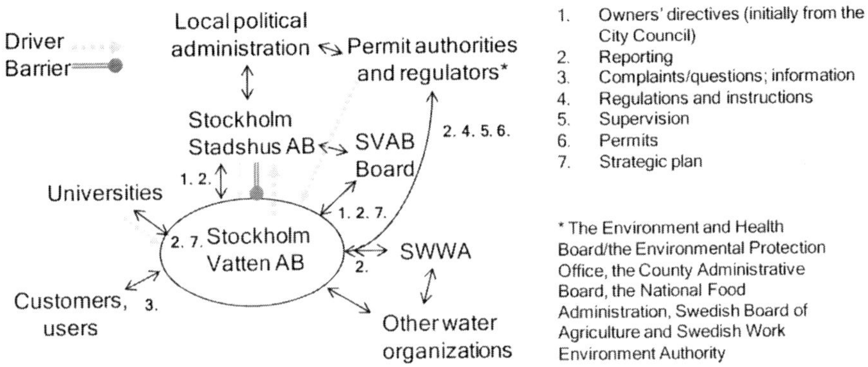

Figure 3. Key actors and information flows in Stockholm water organization.

Customers have not been depicted as drivers, as they were identified as having only an indirect, passive role. In all three organizations, cooperation with universities was mentioned as a potential support (and driver) for developing SDIs. Only in Stockholm, however, did such cooperation actually take place. The influences originating from the municipal context, as seen from the figures, are two-fold in all three cases (depicted in relation to the owner company in the case of SWC). This reflects, on one hand, the fact that the planners in local administrations generally regard SDIs positively; furthermore, the national environmental objectives *should* be followed up at the local level (only in Borås, however, was the local Agenda 21 initiative apparently active). On the other hand, several politicians regarded indicators in a less positive light, acting more or less explicitly as barriers by slowing down the process of developing SDIs. The driver function assigned to permitting authorities and regulators in Figure 1,Figure 2 and Figure 3 arises mainly from the requirement to report to these authorities regarding, for example, nitrogen and phosphorus effluents discharged to receiving waters. These parameters are, like most of the indicators included in the study, not called SDIs, but they do capture some important aspects of sustainability.

4. CONCLUDING DISCUSSION

Of the drivers and barriers identified in the study and summarized above, the key features appear to be attitude to SDIs, trust and communication, the national water sector and regulation, for reasons that will be discussed below.

The attitude to SDIs is central in the sense that it influences and is influenced by several of the other conditions. It consists of two components: (1) indicators as a management tool and (2) sustainable development as a vision and/or normative political goal. The extent to which indicators are appreciated as tools in decision-making, planning and strategy varied considerably between different groups of actors in the study, irrespective of whether sustainable development was considered important or not. In accordance with the theory on SDIs in the introduction, not all actors in the field studies agreed with the use of indicators and the related rational ideal for decision-making. This aspect of indicator use is overlooked in much of the extensive literature on SDIs.

An alternative to SDIs in organizations where the attitude towards them varies from one actor to another could be to concentrate less on the developed SDIs and more on the *process* of developing these indicators and on the possibilities such processes offer in terms of fostering learning and understanding. This appeared to be an often forgotten role of indicators in the utilities studied, only mentioned once in the interviews, the predominant approach to the use of performance indicators being a much more rational one.

Another alternative, or complement, to the rational approach to decision-making would be to focus more on the vision of sustainable development, as was the preferred strategy in Uppsala. The importance of a shared vision—or at least not mutually exclusive visions—of sustainable development is supported by authors such as Rydin *et al.* [34] and Klostermann and Cramer [67]. A precondition for a shared vision of sustainable development appears to be clear communication between the actors involved. Similar findings were made by Wibeck *et al.* [9] in a study of management by objectives in Sweden. Clear communication is facilitated by trust which, in a reciprocal process, can in turn be cultivated by good communication. In the utilities studied, trust and clear communication appeared to be particularly important across organizational boundaries, a conclusion which is consistent with findings in the field of knowledge transfer in organizations (see e.g., Argote and Ingram [68]). The potential drivers, or facilitators, for the development of SDIs are obviously entangled and this study sheds no light on how they relate to one another. *If*, however, SDIs were clearly desired in an organization, trust and communication are aspects that would need to be considered, especially so as SDIs need to "travel" across organizational (system) boundaries to co-ordinate different parts of an organization.

The other component in the attitude towards SDIs is the sense of urgency perceived with regard to the sustainability of present systems. Swedish urban water systems are regarded as sustainable in their present state by many of their actors. Thus, if there is a shared vision of Swedish urban water systems in this instance, it is that they perform fairly well from a sustainability perspective; whether this is valid is not addressed here. What is important here, though, is that there appears to be a correlation between the perception of how sustainable the sector is and the value attached to SDIs: perceived high sustainability of the water system managed correlates to little interest in SDIs (but perceived low sustainability does not necessarily imply a high interest in SDIs, as discussed above apropos the attitude towards indicators as management tools).

The existence of a (more or less) commonly shared vision of the sustainability performance of the sector and the fact that the professional personnel underline the importance of support from and networking within the sector both fit well with the idea of organizational fields in institutional theory [69]. Organizational fields consist of organizations that are institutionally bonded to each other because they perform similar tasks or for some other reason depend on and interact with each other. According to DiMaggio and Powell [68], organizations in the same line of business and within the same organizational field tend to become more similar. The authors identify three mechanisms for such*institutional isomorphic change*, two of which are highly relevant in this context. One stems primarily from professionalization, of which education and professional networks are important aspects, the other from pressure from other organizations, e.g., authorities or other influential organizations. This theory confirms the findings above that the sector itself (through more or less formalized networks and the sector organization) and authorities (through regulation) are important potential drivers for any change in attitude towards sustainable development and a more widespread use of SDIs in urban water management.

Water systems around the world are facing enormous challenges due to population growth, increased use of chemicals and increasing effects of climate change. This is likely to lead to an intensified discussion of what kind of water systems are required to meet these challenges, and possibly (but not necessarily) an increased and widened interest in SDIs to support the process of developing appropriate urban water systems.

ACKNOWLEDGEMENTS

I would like to thank the interviewees for so generously sharing their knowledge and experiences with me, the Swedish Research Council for Environment, Agricultural Sciences and Spatial Planning (FORMAS) for financing the research underlying this paper (project number 25.2/2003-1413), Henrikke Baumann for constructive discussions and recommendations on readings and Roland Clift and two anonymous reviewers for most useful comments that helped improve the paper.

REFERENCES

1. *Water—A Shared Responsibility*; United Nations World Water Development Report No. 2. United Nations World Water Development: Paris, France, 2006. Available online: www.unesco.org/water/wwap (accessed on 4 January 2010).
2. *Revising Indicators of Sustainable Development—Status and Options. Background Paper*; United Nations Division for Sustainable Development: New York, NY, USA, 2006. Available online: http://www.un.org/esa/sustdev/natlinfo/indicators/egmOct06/bgroundPaper.pdf (accessed on 30 September 2009).
3. *Measuring Progress towards a More Sustainable Europe—Sustainable Development Indicators for the European Union—Data 1990–2005*; Office for Official Publications of the European Communities: Luxembourg, 2005.
4. *OECD Environmental Indicators—Development, Measurement and Use. Reference Paper*; OECD Environment Directorate: Paris, France, 2003.
5. Berg, C., Kull, S., Eds.; *Sweden's Environmental Objectives—Buying into a Better Future, de Facto 2006*; Swedish Environmental Objectives Council: Bromma, Sweden, 2006. Available online: http://www.naturvardsverket.se/Documents/publikationer/620-1251-7.pdf (accessed on 4 January 2010).
6. *Sustainable Development Indicators in Your Pocket 2006—An Update of the UK Government Strategy Indicators*; Department for Environment, Food and Rural Affairs: London, UK, 2006.
7. The Boston Indicator Project. *Thinking Globally / Acting Locally—A Regional Wake-Up Call*, 2008. Available

online: http://www.bostonindicators.org/Indicators2008/ (accessed on 30 September 2009).

8. Edvardsson, K. Using goals in environmental management: the Swedish system of environmental objectives. *Environ. Manage.* **2004**, *34*, 170–180. [Google Scholar] [CrossRef] [PubMed]

9. Wibeck, V.; Johansson, M.; Larsson, A.; Åberg, G. Communicative aspects of environmental management by objectives: examples from the Swedish context. *Environ. Manage.* **2006**, *37*, 461–469. [Google Scholar] [CrossRef]

10. Global Reporting Initiative (GRI). Available online: http://www.globalreporting.org/Home(accessed on 30 September 2009).

11. Balkema, A.J.; Preisig, H.A.; Otterpohl, R.; Lambert, F.J.D. Indicators for the sustainability assessment of wastewater treatment systems. *Urban Water* **2002**, *4*, 153–161.

12. Coulibaly, H.D.; Rodriguez, M.J. Development of performance indicators for small Quebec drinking water utilities. *J. Environ. Manage.* **2004**, *73*, 243–255.

13. Foxon, T.J.; McIlkenny, G.; Gilmour, D.; Oltean-Dumbrava, C.; Souter, N.; Ashley, R.; Butler, D.; Pearson, P.; Jowitt, P.; Moir, J. Sustainability criteria for decision support in the UK water industry.*J. Environ. Plann. Man.* **2002**, *45*, 285–301.

14. Hellström, D.; Jeppsson, U.; Kärrman, E. A framework for systems analysis of sustainable urban water management. *Environ. Impact Asses.* **2000**, *20*, 311–321.

15. Lundin, M.; Molander, S.; Morrison, G.M. A set of indicators for the assessment of temporal variations in the sustainability of sanitary systems. *Water Sci. Technol.* **1999**, *39*, 235–242.

16. Mactavish, A. *Towards Sustainability*; UK Water Industry Research: London, UK, 2003.

17. Malmqvist, P.A.; Palmquist, H. Decision support tools for urban water and wastewater systems—focusing on hazardous flows assessment. *Water Sci. Technol.* **2005**, *51*, 41–49.

18. Sahely, H.R.; Kennedy, C.A.; Adams, B.J. Developing sustainability criteria for urban infrastructure systems. *Can. J. Civil Eng.* **2005**, *32*, 72–85.

19. Sundberg, C.; Svensson, G.; Söderberg, H. Re-framing the assessment of sustainable stormwater systems. *Clean Technol. Environ.* **2004**, *6*, 120–127.

20. Palme, U.; Lundin, M.; Tillman, A.-M.; Molander, S. Sustainable development indicators for wastewater systems—researchers and indicator users in a co-operative case study. *Resour. Conserv. Recy.* **2005**, *43*, 293–311.

21. McCool, S.F.; Stankey, G.H. Indicators of sustainability: challenges and opportunities at the interface of science and policy. *Environ. Manage.* **2004**, *33*, 294–305.

22. de Kruijf, H.A.M.; van Vuuren, D.P. Following sustainable development in relation to the North-South Dialogue: ecosystem health and sustainability indicators. *Ecotox. Environ. Safety* **1998**, *40*, 4–14.

23. Verbruggen, H.; Kuik, O. Indicators of sustainable development: an overview. In *In Search of Indicators of Sustainable Development*; Kuik, O., Verbruggen, H., Eds.; Kluwer Academic Publishers: Dordrecht, The Netherlands, 1991; pp. 1–6.

24. Hezri, A.A.; Dovers, S.R. Sustainability indicators, policy and governance: issues for ecological economics. *Ecol. Econ.* **2006**, *60*, 86.

25. Innes, J.E.; Booher, D.E. Indicators for sustainable communities: a strategy building on complexity theory and distributed intelligence. *Plan. Theory Pract.* **2000**, *1*, 173–186.

26. Meadows, D. *Indicators and Information Systems for Sustainable Development*; The Sustainability Institute: Hartland Four Corners, VT, USA, 1998.

27. Palme, U.; Tillman, A.-M. Sustainable development indicators: how are they used in Swedish water utilities? *J. Clean. Prod.* **2008**, *16*, 1346–1357.

28. Hatch, M.J. *Organisationsteori. Moderna, Symboliska och Postmoderna Perspektiv*; Swedish edition; Studentlitteratur: Lund, Sweden, 2002.

29. Thompson, J.D.; Tuden, A. Strategies, structures and processes of organizational decision. In*Comparative Studies in Administration*; Thompson, J.D., Hammond, P.B., Hawkes, R.W., Junker, B.H., Tuden, A., Eds.; University of Pittsburgh Press: Pittsburgh, PA, USA, 1959; pp. 195–216.

30. Thompson, J.D. *Organizations in Action*; McGraw-Hill: New York, NY, USA, 1967.

31. Chaffee, E.E. Three models of strategy. *Acad. Manage. Rev.* **1985**, *10*, 89–98.

32. Rich, R.F.; Oh, C.H. Rationality and use of information in policy decisions: a search for alternatives. *Sci. Commun.* **2000**, *22*, 173–211.

33. Mintzberg, H. The design school: reconsidering the basic premises of strategic management.*Strateg. Manage. J.* **1990**, *11*, 171–195.

34. Rydin, Y.; Holman, N.; Hands, V.; Sommer, F. Incorporating sustainable development concerns into an urban regenation project: how politics can defeat procedures.*J. Environ. Plann. Man.* **2003**,*46*, 545–561.

35. Mitchell, G. Problems and fundamentals of sustainable development indicators. *Sustain. Dev.***1996**, *4*, 1–11.

36. *Facts on Water Supply and Sanitation in Sweden*; The Swedish Water and Wastewater Association: Stockholm, Sweden, 2000.

37. Wilderer, P.A. Applying sustainable water management concepts in rural and urban areas: some thoughts about reasons, means and needs. *Water Sci. Technol.* **2004**, *49*, 7–16.

38. *The Swedish Environmental Objectives—Interim Targets and Action Strategies*; Government Bill 2000/01:130. The Swedish Government: Stockholm, Sweden, 2001. Available online: http://www.regeringen.se/content/1/c4/11/97/7808aab9.pdf (accessed on 4 January 2010).

39. Yin, R.K. *Case Study Research: Design and Methods*, 3rd ed.; Sage Publications: Thousand Oaks, CA, USA, 2003.

40. Arbnor, I.; Bjerke, B. *Företagsekonomisk Metodlära*; Studentlitteratur: Lund, Sweden, 1977.

41. Miles, M.B.; Huberman, M.A. *Qualitative Data Analysis*; SAGE Publications: Thousand Oaks, CA, USA, 1994.

42. Palme, U. *Indikatorer för Hållbar Utveckling av Urbana VA-System? Fallstudie Uppsala (Indicators for Sustainable Development of Urban Water Systems? Field Study Uppsala),Environmental Systems Analysis*; Chalmers University of Technology: Göteborg, Sweden, 2005.

43. Palme, U. *Indikatorer för Hållbar Utveckling av Urbana VA-System? Fältstudie Borås (Indicators for Sustainable Development of Urban Water Systems? Field study Borås),Environmental Systems Analysis*; Chalmers University of Technology: Göteborg. Sweden, 2005.

44. Palme, U. *Indikatorer för Hållbar Utveckling av Urbana VA-System? Fältstudie Stockholm Vatten (Indicators for Sustainable Development of Urban Water Systems? Field study Stockholm Water), Environmental Systems Analysis*; Chalmers University of Technology: Göteborg, Sweden, 2006.

45. Denscombe, M. *The Good Research Guide,*, 2nd ed.; Open University Press: Philadelphia, PA, USA, 2003.

46. Kvale, S. *Den Kvalitativa Forskningsintervjun*; Studentlitteratur: Lund, Sweden, 1997.

47. *Miljöredovisning*; Uppsala Kommun: Uppsala, Sweden, 2000. Available online: http://www.uppsala.se/Upload/Dokumentarkiv/Externt/Dokument/Om_kommunen/Miljoarbete/miljoredovsining2000.pdf (accessed on 4 January 2010).

48. *Miljömål för Borås*; Borås Agenda 21: Borås, Sweden, 2003. Available online: http://www.boras.se/download/18.633e5e10039748abd7fff53325/nr70.pdf (accessed on 4 January 2010).

49. Borås Water and Sanitation Department. *Verksamhetsmål för VA-Verksamheten i Borås (Preliminary Goal Document for the Water Organization in Borås)*; Borås Stad: Borås, Sweden, 2004.

50. Frank, I.; Henriksson, B.; Olsson, K.; Reinius, L.-G.; Sjöberg, G.-M.; Värnhed, B. *Stockholm Vatten AB Nyckeltal 1991–2000 (Stockholm Water Company Indicators 1991–2000)*; Stockholm Vatten: Stockholm, Sweden, 2001.

51. Stockholm Water Company. *Nyckeltal 1999–2008 (Indicators 1999–2008)*; Stockholm Vatten: Stockholm, Sweden, 2008.

52. Stockholm Stadshus AB. *Uppföljning av Kommunfullmäktiges Övergripande Inriktningsmål 2004. Bilaga till Bokslut för Koncernen*; Stockholms Stad: Stockholm, Sweden, 2005.

53. Stockholm Stadshus AB. *Uppföljning av Kommunfullmäktiges Övergripande Inriktningsmål 2004. Bilaga till Bokslut för Koncernen. (Follow-up of the Municipal Council's Overall*

Objectives for 2004. Appendix to the Closing of Accounts for the Group); Stockholms Stad: Stockholm, Sweden, 2004. (In Swedish)

54. Frank, I.; Hådell, C.; Lindblom, L.; Snith, Å.; Osvald, K.; Sahlin, E.; Sjöberg, G.-M.; Värnhed, B.*Nyckeltal 1995–2004 (Indicators 1995–2004)*; Stockholm Vatten: Stockholm, Sweden, 2005.

55. Frank, I.; Hådell, C.; Hellström, D.; Dahlberg, U.; Finnson, A.; Behdjou, M.; Hult, P.; Sjöberg, G.-M.; Tropp, M.; Lindblom, L. *Stockholm Vatten AB Nyckeltal 1996–2005 (Stockholm Water Company Indicators 1996–2005)*; Stockholm Vatten: Stockholm, Sweden, 2006.

56. *ISO 14001 and beyond: Environmental Management Systems and beyond*; Sheldon, C., Ed.; Greenleaf Publishing: Sheffield, UK, 1997.

57. Brown, R. Local institutional development and organizational change for advancing sustainable urban water futures. *Environ. Manage.* **2008**, *41*, 221–233.

58. Earle, T.C.; Siegrist, M. Morality information, performance information, and the distinction between trust and confidence1. *J. Appl. Soc. Psychol.* **2006**, *36*, 383–416.

59. Evans, B.; Joas, M.; Sundback, S.; Theobald, K. Governing local sustainability. *J. Environ. Plann. Man.* **2006**, *49*, 849–867.

60. McNie, E.C. Reconciling the supply of scientific information with user demands: an analysis of the problem and review of the literature. *Environ. Sci. Policy* **2007**, *10*, 17–38.

61. Hood, C. The "new public management" in the 1980s: variations on a theme. *Account. Org. Soc.* **1995**, *20*, 93–109.

62. Montin, S. *Moderna Kommuner (Modern Municipalities)*; Liber: Malmö, Sweden, 2002.

63. von Malmborg, F. Environmental management systems: what's in it for local authorities? *J. Environ. Policy Plan.* **2003**, *5*, 3–21.

64. Helland, B.; Adamsson, J. Performance indicators: benchmarking between six cities in Scandinavia. *J. Water Serv. Res. Tec.* **1998**, *47*, 284–288.

65. *Energi 2005*; Swedish Water and Wastewater Association: Stockholm, Sweden, 2006. Available online: http://www.svensktvatten.se/Templates/FileArchive1.aspx?PageID=e20309eb-fb49-4ae5-84ea-af689bea63bd (accessed on 30 September 2009).

66. Dominguez, D.; Gujer, W. Evolution of a wastewater treatment plant challenges traditional design concepts. *Water Res.* **2006**, *40*, 1389–1396.

67. Klostermann, J.E.M.; Cramer, J. Social construction of sustainability in water companies in the Dutch coastal zone. *J. Clean. Prod.* **2007**, *15*, 1573–1584.

68. Argote, L.; Ingram, P. Knowledge transfer: a basis for competitive advantage in firms. *Organ. Behav. Hum. Dec.* **2000**, *82*, 150–169.

69. DiMaggio, P.J.; Powell, W.W. The iron cage revisited: institutional isomorphism and collective rationality in organizational fields. *Am. Sociol. Rev.* **1983**, *48*, 147–160.

4
CHAPTER

Waste Water Management Systems

Jelenka Savković-Stevanović[1]

[1] Department of Chemical Engineering, Faculty of Technology and Metallurgy, Belgrade University, Serbia

1. INTRODUCTION

As mankind eventually adopted a more settled, non-nomadic way of life, people become increasingly involved in the technical aspects of water. To a large extent, the primary concerns in the beginning were utilization and improvement of existing water resources, together with protection against the hazards and potential harm associated with uncontrolled natural water. It was only toward the end of the nineteenth century that wastewater become an issue in science, technology, and legislation, specifically, its production and treatment, in terms of both municipal and industrial sources [1]-[5].

As early as 4500 years ago the first prerequisites were met for urban and agricultural water management. This encompassed irrigation and drainage systems, canals, and sewage facilities. Even so, the treatment of waste water in formal waste treatment plants by means of the microbial degradation of wastewater components was reported for the first time 1892. It was municipal wastewater, including that from artisans, craftsman, and small factories, typical for larger cities of that time. Much earlier, in Greco- Roman times extensive facilities were erected and maintained for supplying drinking water to cities. Waste water in those days presented no major problems, and sewage systems were regarded less as a means of collecting water for reuse than as way of draining off potential sources of hazard and preventing pollution of the streets, with the attendant risk of a spread of vermin and epidemics.

Development of the organized utilization of water as an essential resource for human beings, animals, and plants led to further technical strides, such as dams against flooding or for storage purposes, waterways designed for transport, and harbors on the sea coast and along inland waterways. Problems of wastewater arose gradually during the same period in conjunction with the increase in urban population as the natural self purification capacity of surfaces waters proved no longer able to keep pace with development. Risks related to groundwater contamination are associated not only with emissions in the form of wastewater, the ground - air cycle plays a role as well via atmospheric deposition. The state of a particular body of water can be described by a set of code numbers, but the core of the problem is in fact a sum of all the processes leading to the observed state. In this case, the essential element is the kinetics of two opposing processes the rate of pollution, and the rate of cleansing. Each of these is in turn a combination of natural and anthropogenic phenomena applicable to the site in question. Most of the problems are derived not from absolute numbers, but rather from population densities, production densities, or productivities, in the various urban centers of the industrialized world, all of which actually have access to an adequate supply of natural water.

In the terminology of water economics, consumption of water refers to a loss of quantity, not a decrease in quality. In this sense, consumption represents that part of the water supply that is lost in the course of use, primarily through evaporation. This fraction of the water is permanently withdrawn, at least from the local water cycles, and is thus no longer available for further utilization, so it must be replenished with water from precipitation, springs, or wells.

As industrialization proceeded, two unique characteristics of water acquired rapidly increasing importance, its high specific heat capacity, and its rather high solvent power with respect to many inorganic and some organic substances. The consequences of these factors in the context of the production and disposal of wastewater are quite different, however. Water that is intended to serve as a heat reservoir, cooling agent, steam source, must be cleaned before use in order to prevent corrosion and erosion in turbines and heat exchangers, and it is subsequently returned to the environment in a purified state, albeit at a higher temperature. On the other hand, water in its function as a reaction medium, or even a reaction partner, has now developed into the most significant source of wastewater in industry. Entire branches of manufacturing are based on production processes carried out in the aqueous phase, where water is used as a solvent, dispersing agent, transport medium, and reagent. This is perhaps most evident in the case of breweries, in sugar, paper, and pulp factories, in dye works, tanneries, and the like where the teal problem is one not only of the actual content of the wastewater, but also its quantity [6].

Water in drawn by industry from many different sources. It may be taken directly from a river, a lake, a well, or from a privately impounded supply, or it may be obtained from a neighboring municipality. Both the amount drawn by the industry and the degree of treatment accorded the water so withdrawn varies widely from industry to industry and from plant to plant. The quality of treatment may vary considerably within a given plant depending upon the particular uses to which the water is put. the amounts of water withdrawn by various industries for different uses, and the quality of water s that have been used by different industries before being subjected to various degree of treatment are varied.

Water has been used in abundant quantities by chemical, petrochemical, petroleum refining and other process industries. However, in recent years, the increased cost of wastewater treatment to meet environmental requirements and the scarcity of less expensive industrial water have provided process industries with strong incentive to minimize the amount of water consumption and wastewater discharge. The major concern is to emphasize the importance of water reuse and a number of effort have been made towards achieving the goal of extensive water reuse in various process industries [3].

There have been presented many ideas for wastewater recovery and reuse in the in industries[5]-[9]. These paper have exclusively described wastewater treating systems for the realization of zero discharge. As for the optimal design methods for wastewater treating systems, several attempts have also been made by using system approaches. Much information on the optimization studies on process units for waste water treatment can be acquired from this survey [10]-[50]. In addition, a method of utilizing the system structure variables is considered to be useful to eliminate difficulties due to combinatorial problems. The studies presented so far, however, only cover wastewater treating systems. The amount of wastewater was given beforehand and its reduction was not taken into consideration. As far as the authors know, the optimal design problem including water reuse for the total system consisting of water-using system and wastewater -treating system has not yet been solved.

2. BASIC PRINCIPLE

In the last decade, a number of studies, on wastewater reuse or optimal designs of waste water treating systems have been presented. Though those studies have received much attention, they have been carried out exclusively on wastewater treating systems without paying attention to water using systems. However, the authors extensive survey on the present status of water use in a industry has shown that there is enough room to reduce a large amount of both fresh water and wastewater. The reduction can be accomplished by optimizing water allocation in a total system consisting of water using units and wastewater treating units. The problem of maximizing water reuse can be considered as a problem of optimizing water allocation in a total system. Furthermore, the problem of determining a system structure is defined as a parameter optimization problem by employing structure variables. Due to the approach, the difficulties associated with combinatorial problems are resolved.

For the year 2000, a global balance of quantities and fluxes shows an overall water supply of 2500 $km^3$and water demand 6000 km^3, representing 24% of the directly usable supply [6]. When this allocated among the major consumers and account is taken of the corresponding levels of specific water consumption (evaporation), several trends with regard to quantities and types of wastewater are discernible. (Table 1)

TABLE 1. Types of wastewater.

Consumption category	Percentage of total demand (6000 km^3)	Consumption (evaporation) as a percentage of demand
Domestic	8	20-30
Industry	29	15-20
Agriculture	59	75
Storage losses	4	100

Water circulation through the atmosphere contains 13000 km^3,ca. 0.02% of the overall liquid, global water reserve of 1.3 x 10^9 km^3. The annual quantity subject to evaporation, which is equal to the annual amount of precipitation, is estimated at 475000 km^3. This corresponds to a 35-fold annual turnover of the atmospheric content, which means water exchange between the atmosphere and the surface of the earth is complete every 9.5

days. In some cases local circumstances lead to considerable deviations from these global values.

Waste water treatment becomes especially important in times of water scarcity. This is particularly true when agriculture, domestic water needs, and industry find themselves in vigorous competition. the greatest increase by far is anticipated for agriculture assuming for the year 2000 a world population 6-6.5 x 10^9 (Fig.1).

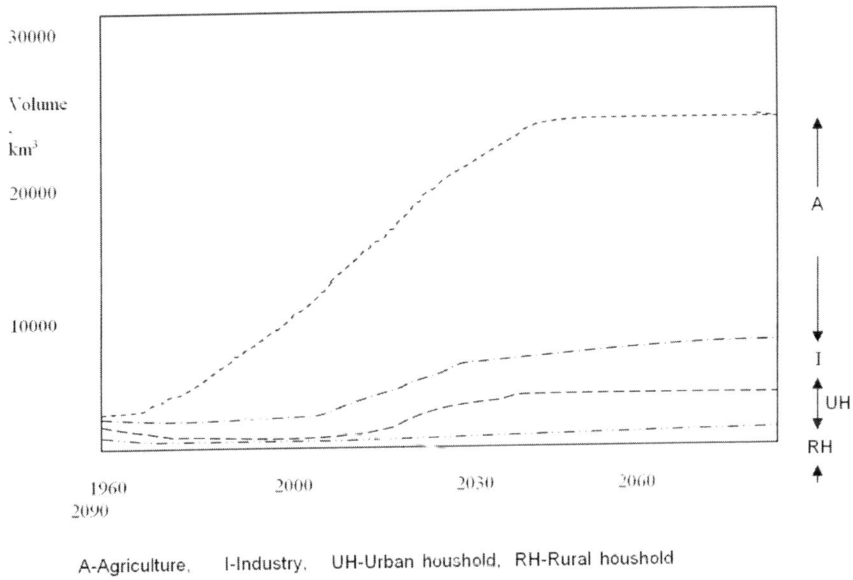

FIGURE 1. The increasing worldwide demand for water.

Water as regarded as chief raw material problem of the future, elevating wastewater treatment to the status of a recycling technology. This involves substance protection in the sense of sustainable development, together with the maintenance of an adequate supply of drinking water, an emphasis that goes beyond the earlier concern directed almost exclusively toward environmental protection, especially the protection of natural waters.

Each of the processes described below has its place in the broad spectrum of technical possibilities. The question of what is the best process should thus be replaced by a search for the most suitable process in a particular circumstance, taking fully into account the nature of certain definable problem cases. Modern technological development of wastewater treatment has occurred largely in and with the aid of the chemical industry. However, the waste water problem and its treatment is of interest not only to this particular aspect of industrialized society. Numerous other branches, often to an even greater extent (Fig.2).

There is no shortage today of diverse international experience in the construction and operation of waste- treatment plants. In fact, in some places this has developed into its own separate branch of process engineering. It is still worth noting with respect to terminology, however, that some of the reasoning applied to wastewater concepts and standards of evaluation has been borrowed from neighboring disciplines, especially biology.

3. INDUSTRIAL WATER TREATING SYSTEMS

The authors have carried out an expensive study on the present status of water use in a typical industry. As the result it has been shown that there enough room to reduce a large amount of wastewater by maximizing water reuse and waste water recovery. Further increases in the efficiency of water use can be expected by the change of process conditions. In solving such large complex problems, it has been found more methods.

Water is drawn by industry from many different sources. It may be taken directly from a river, a lake a well, or from a privately impounded apply, or it may be obtained from a neighboring municipality. Both the amount drawn by the industry and the degree of treatment accorded the water so withdrawn varies widely from industry to industry and from plant to plant [10]-[36]. The quality of treatment may vary considerably within a given plant depending upon the particular uses to which the water is put. Table 2shows the amounts of water withdrawn by various industries for different uses[37]-[40].

In the chemical industry, economic factors usually dictate the inclusion of a wastewater separation system. Thus, wastewater that is not in need of treatment, clean water, especially cooling water, is separated from that which does require treatment. Clean water can be discharged directly into the receiving stream. If the wastewater requiring treatment fails to meet the quality specifications for biological waste treatment it must first be subjected to decentralized chemical-physical pretreatment, after which it can be fed into the central wastewater treatment plant for purification as shown in Fig.2.

Specific wastewater loads can be reduced or even avoided through measures of the type recommended in conjunction with process integrated environmental protection.

The entire wastewater regime is also reflected in the approach taken to wastewater decision making. For each type of wastewater, all the following questions must be rigorously addressed to ensure consistent and disposal (Fig.3).

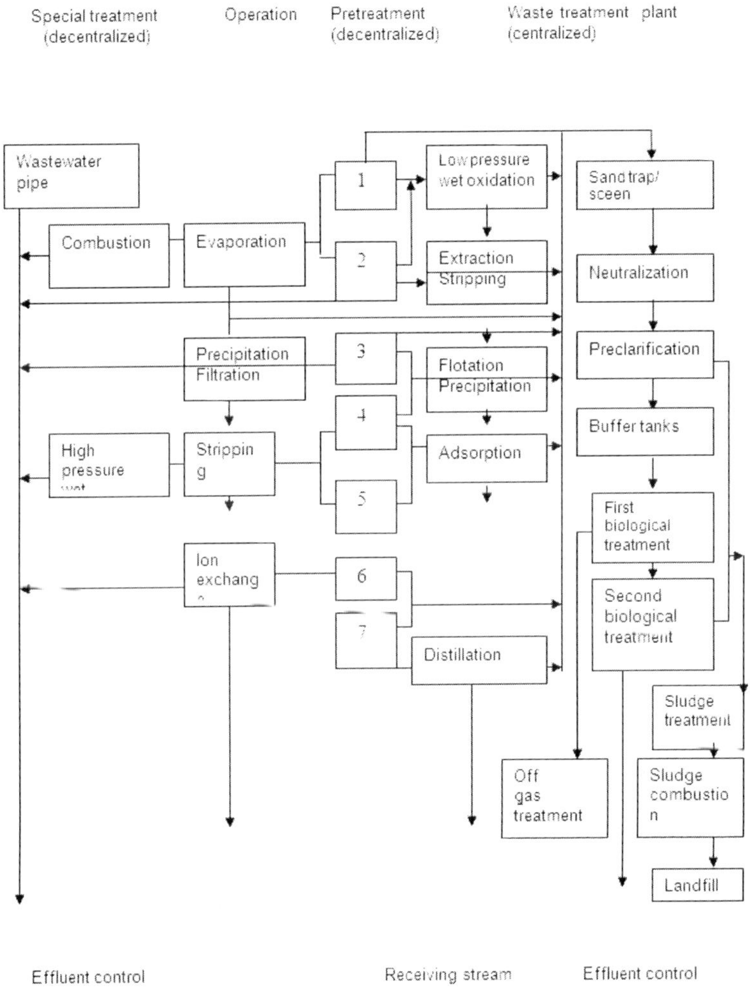

FIGURE 2. Decentralized and centralized aspects of waste water treatment.

Can be amount and contamination level of the waste water be reduced or even eliminated by process integrated means?

Does all the wastewater in question in fact require treatment?

Is the waste water suitable in its present form for biological treatment, or should it be subjected to decentralized pretreatment?

These considerations apply, for companies that have their own central wastewater treatment plants, and indirect discharge, for companies that dispose of their wastewater via a public wastewater facility.

Suspended solids removal, particularly of the coarser materials from 5μ up, such as sand and heavy silt may be removed in sedimentation basins. Such basins usually serve a dual purpose, preliminary removal of suspended solids, and storage to balance variations in supply with the relatively constant demand of the plant processes. In these basins, detention time is measured in days, the amount depending upon the likelihood of interruption or reduction in the supply. A 30th-day detention is not uncommon in some circumstances. Particles smaller than 1μ are generally not affected by the detention. The effect of continued aeration and sunlight in oxidizing organic peptizing substances may cause a certain amount of flocculation. Algal growth, particularly of the free floating type, occurs in warmer climates and may indeed contribute to the total turbidity emerging from the basin. Control of algal growth is usually accomplished by the addition of copper sulfate sprayed in aqueous solution on the water surface from a boat or spread by solution from solid material in burlap bags towed behind power boats that traverse the surface of the reservoir in a pattern. In the warmer climates, addition of copper sulfate every several months in the amount of one ppm, based on the top one foot of water, may be employed. In some exceptional circumstances in very arid regions with short water supply, evaporation control may be practiced by the addition of fatty alcohols which form a monolayer on the surface.

The degree of clarification applied to the water from the source of supply is governed both the intended use of the water and by the level of turbidity that is to be removed. Two processes, singly or in combination, are generally employed for clarification. The first of these is filtration where turbidities are generally less than 50 ppm, and clarification can be accomplished simple by passing the water through a filter. Filters may be of three configurations, the most common being a single granular medium such as sand, typically of effective particle size of 0.4 mm at flow rates ranging from 1 to 8 gal/minft^{2}.

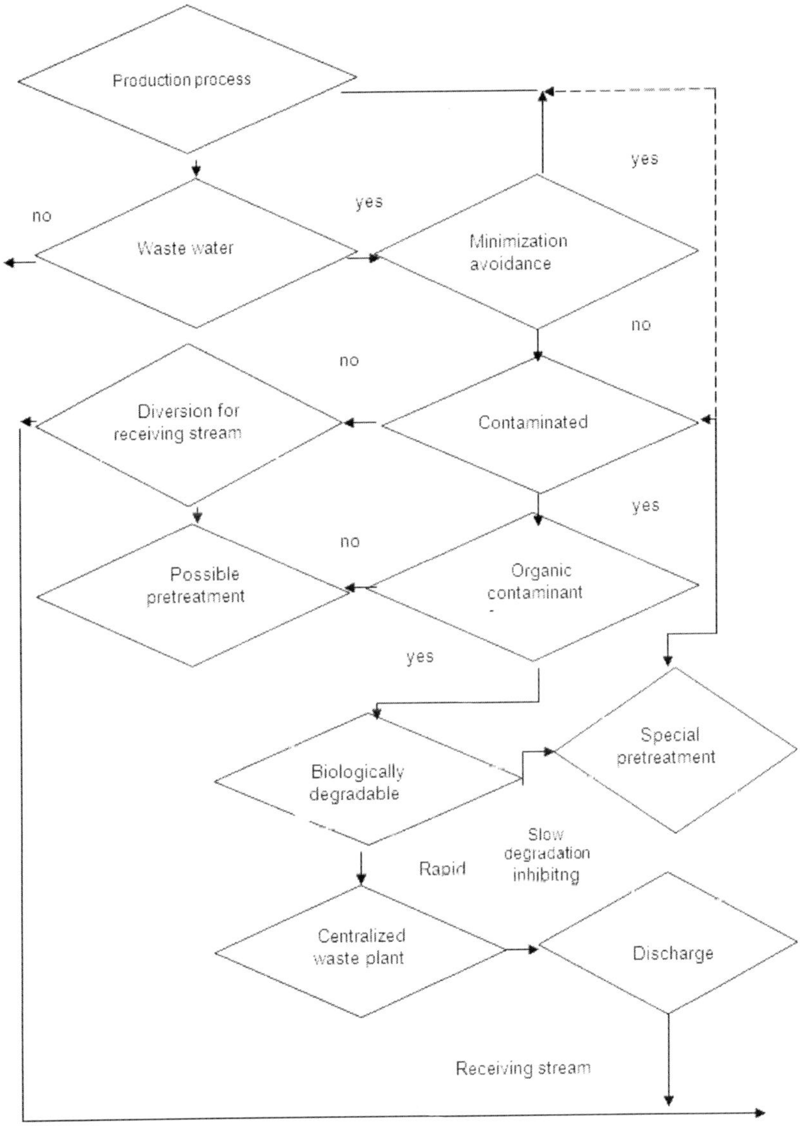

FIGURE 3. Decision diagram dealing with for waste water formation, avoidance, separation and treatment.

A small amount of a coagulant such as alum may be added ahead of the filter amounts varying from 5 to 15ppm. Such rapid sand filters may either be operated by gravity, relying solely on a head of water over the filter medium to force the water through, or they may be completely enclosed in a cylindrical tank with pump pressure used to force the water through the medium. Pressure drops typically range from 1 to a maximum of 8 ft of water. The depth of bed employed is commonly from 2 to 3 ft. Finely divided anthracite coal of effective particles size of 0.6 mm may be used instead a sand. When anthracite is used, the addition of coagulation chemicals is usually required. The coagulation chemicals added serve the purpose of agglomerating the colloidal dispersed solids and aid in their adherence to the filter media and hence their removed. Clarification efficiencies measured as the ratio of suspended solids entering are commonly 0.90-0.99. The sand or anthracite coal in earlier designs was supported on a bed of graded gravel, but in current practice is more commonly supported directly on the filter bottom which is provided with strainers sufficiently fine to prevent sand or anthracite from passing through.

Previously most filters, both pressure and gravity, were of a single medium, sand or anthracite. Currently, many new installations are being designed with a mixed media or graded density filter. In this case, filter media of different types such as sand and anthracite are employed together, with the anthracite, being the more coarse and lower density medium, appearing on the top of the filter, 49 mm (20 in) of 1.0 mm anthracite coal may be placed on top of 14.7 mm 6 in (6 in) of 0.4 mm sand. When the filter backwashed to remove the suspended impurities that accumulate during the run, the less dense medium, the coal, is washed to the top by the upward flow of water and the heavier medium, the sand, even though finer, remains on the bottom. The suspended solids that have been removed, being much lighter than, either medium and in flocculated form, are carried out by the up flowing water and washed to waste. In both types of filters, single medium and mixed media, the amount of backwash water required is approximately three percent of the total throughput of the filter during a run, the end of which is controlled by a given limit on the quality of the effluent. The advantages of the mixed media filter are that the coarser material on top causes removal of a large percentage of the suspended solids in the entering water and allows the solids removed to accumulate in the depth of the bed rather than forming a mat on the top. The remaining amount of suspended solids is removed on the finer media. The net results is an ability to handle an influent water with a much higher suspended solids content, and to process a greater quantity at higher flow rates without deterioration of effluent quality than is possible with a single medium filter. Where mixed media filters are employed, flow rates vary from 3 to 6 gal/minft^2. Total head loss is limited to approximately results in a penetration of suspended solids through the filter medium into the effluent.

TABLE 2. Industrial plant and thermal-electric-plant (Water intake, Reuse, and Consumption, 1964.

Industrial Group	Water intake, billion (gal /year)[b] -cooling and condensing	Water intake, billion gal/ year -boiler feed sanitary service etc.	Water intake, billion gal/ year -proc Ss	Water intake, billion gal/ year -total	Water intake, billion gal/ year -water recycled	Water intake, billion gal/ year -gross water use, including recycling	Water intake, billion gal/ year -water consumed	Water intake, billion gal/ year -water discharged
Food and kindred products	392	104	264	760	520	1280	72	688
Textile mill products	24	17	106	147	163	810	13	134
Lumber and wood products	71	24	56	151	66	217	28	123
Paper and allied products	607	120	1344	2071	3045	6016	129	1942
Chemicals and allied products	3120	202	564	3880	3688	7574	227	3650
Petroleum and coal products	1212	99	88	1399	4768	6162	81	1318
Leather and leather products	1	1	14	16	2	18	1	15
Primary and metal industry	3387	195	995	4578	2200	6778	266	4312
subtotal	8814	762	3432	13008	15347	28355	817	12191
other industries	571	197	271	1039	1207	2246	71	968
Total industry	9385	959	3703	14047	16554	30601	888	13159
Thermal Electric Plants	34849	c		34849	5815	40665	68	34781
Total	44234	959[c]	3703	48896	22869	71265	956	44940

[a] [37]).a Source census of manufactures [37], b gal means British gallon 1gal= 4.54 L, c Boiler feed water use by thermal electric plants is estimated to be equivalent to sanitary service, in industrial plants etc., d Total boiler feed water(excluding sanitary service in industrial plants).

4. ADSORPTIVE, CHEMICAL AND INCINERATION SYSTEMS IN WASTER WATER TREATMENT

Adsorption processes have been successfully applied in the chemical industry for the purification of water generally, as well as for various solutions and individual wastewater streams.

In the chemical, physical, and biological purification of wastewater dissolved constituents are eliminated not only by use of special absorbents, but also in many cases by adsorption on the surface of undissolved substances already present. Elimination occurring by the latter mechanism cannot be examined separately, so it is simply described to the treatment process as a whole.

The deliberate application of adsorption is always in competition with other chemical, physical, and biological approaches to the purification of wastewater. Ecological and economic considerations are decisive in the choose of any particular process or process combination [10]. The question as to whether adsorption is technically and economically feasible, either alone or in combination with other processes, must be examined separately in each individual case. To this end it is necessary to consider the adsorption phenomenon itself in conjunction with the preparation and regeneration of the adsorbent, since from a process engineering stand point the two processes constitute a single entity. In addition, reuse or disposal of material arising in the course of the process should be taken into account as appropriate.

Adsorption causes dissolved organic wastewater constituents to accumulate at the surface of one or more adsorbents. Adsorption must there fore be regarded as a physical concentration process taking place at a liquid solid phase boundary, one that competes with other concentration processes occurring at liquid-liquid [11]-[13] (extraction or membrane processes) and liquid-gas boundaries (evaporation, distillation, stripping)[34]. In the unified examination that follows, it was necessary that some attention also be devoted to such oxidative processes as aerobic biological treatment, chemical oxidation, wet oxidation, and combustion, treated elsewhere in this contribution in detail.

In general, it may be assumed that adsorption processes have already established themselves as valuable for dealing with fairly small wastewater streams containing low concentrations of adsorbable substances. Adsorption is especially common in decentralized waste water treatment, both alone and in combination with other processes.

Dissolved waste water constituents are potentially subject to attachment at the surfaces of solids. An accumulation of this type is known as adsorption, and it is attributable mainly to van der Waals forces, particularly dipole-dipole interaction, though columbic forces also often play an important role. The fixing of dissolved substance to a sorbent can sometimes lead to an enrichment factor of 10^5 or greater. In most cases the adsorbent–adsorbate mixture must then be worked up in a second processing step. Regeneration of a contaminated sorbent frequently takes advantage of the reverse counterpart to adsorption-desorption.

The equilibrium state that is established after a sufficient amount of time has elapsed defines the equilibrium concentration of a solute in a liquid and the loading of an adsorbent. This state can be described in terms of so called adsorption isotherms. An isotherm is applicable only to a specific, defined temperature, although the influence of temperature is very small in liquid-solid systems in contrast to gas-solid systems. The curves themselves are usually determined empirically, but their shape can also be described mathematically.

One of the first attempts at developing an isotherm equation was reported by Langmuir, who was concerned originally with the adsorption of gases, basing his work on methods of statistical thermodynamics. The Langmuir equation assumes the presence of a adsorbent surface, and it is therefore valid only to the point of a monomolecular covering of the relevant surface. In general, Langmuir homogenous isotherms are described by the equation:

$$X = X_m \frac{c}{b+c} \tag{1}$$

where X is equilibrium loading, X_m is loading for monomolecular coverage, c is residual solution concentration remaining after establishment of the adsorption equilibrium, and *b* is *a* constant.

The isotherm equation proposed by Freundlich as early as Langmuir was established empirically. Because of its relatively simple form it is an excellent device for describing adsorption from an aqueous solution:

$$\frac{X}{M} = kc^n \tag{2}$$

where X is amount of adsorbed substance, M is weight of adsorbent used, c is residual concentration remaining after establishment of the adsorption equilibrium, and k and n are constants specific to a given adsorbent-adsorbate mixture.

Plotting this equation in log-log function, $\log(X/M) = \log k + n\log c$, produces a straight line, which greatly facilitates practical application of the method (Fig.4 and Fig. 5).

The Langmuir equation is clearly limited to low degree of loading, whereas in the case of the Freundlich equation there is no formal concentration-dependent limitation to the expression's validity. In practice, however a flattening of the Freundlich isotherm does become evident with increasing concentration, which means that the Freundlich exponent (n) increases with decreasing concentration. Thus rigorous application of the Freundlich equation is limited to a rather narrow concentration range.

Multicomponent mixtures are the rule in practical waste water treatment, and these can be characterized only by such cumulative parameters. Individual substances included in a collective analysis compete for occupation of the available adsorption sites.

Since different substances usually have very different adsorption behaviors, the Freundlich isotherms actually obtained in such cases are curved, or even bent at sharp angles as shown in Fig.5.

Despite the complications created by isotherms whose course is not straight, the Freundlich isotherm method is frequently used as a way of rapidly acquiring information regarding absorbability on adsorbents. Moreover, such isotherms make it possible to estimate adsorbent consumption rates as well as the achievability of particular concentration targets.

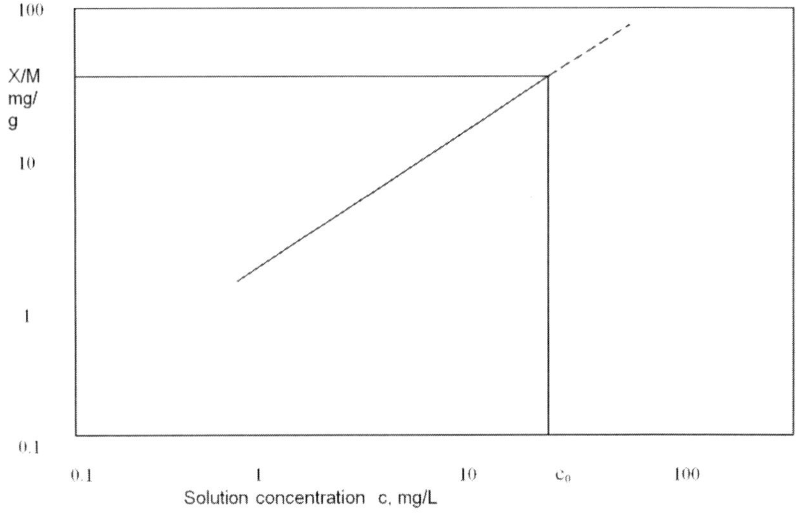

FIGURE 4. A well behaved Freundlich isotherm.

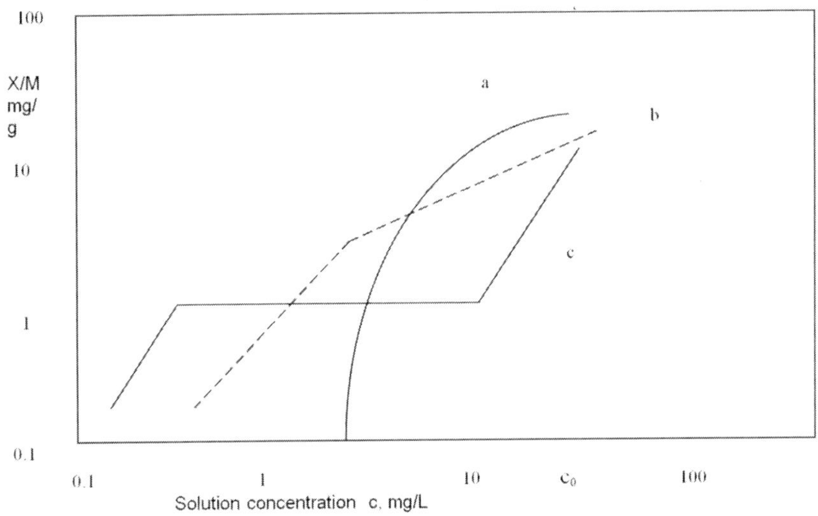

FIGURE 5. Typical non-linear Freundlich isotherms (a,b,c) obtained with different mixtures of solute.

In the last twenty years special processes have been developed to utilize chemical reactions directly for the degradation of water contaminants. Some have been developed to the point of technical maturity, and without exception these are oxidative processes. The methods for the oxidative and reductive elimination of wastewater components are old.

The active oxidizing agents are oxygen containing radicals in the vast majority of cases the radical H-O. The oxidizing agents actually introduced are usually air, oxygen, hydrogen peroxide, or ozone. Process engineering characteristics distinguishing the individual methods vary quite considerably. What differentiates one process from another. and distinguishes all of these approaches from the microbial route, is a consequence largely of the nature of the wastewater itself and the extent to which its components should or must be degraded. It should also be noted that advantage is sometimes taken of several either direct chemical or biochemical methods, as well as combinations of the two together.

The generally more drastic reaction conditions or more aggressive reactants associated with a purely chemical method, and the often higher degradation rates achieved, lead easily to the impression that direct chemical methods are inevitably superior to biochemical methods. This is not necessarily the case, however. Material degradation is the first and

foremost a kinetic problem, and reaction mechanisms and catalysis play decisive roles. To describe a substance as "not degradable" is basically an unwarranted oversimplification, because it implies that "degradability" is a fixed property of a substance, whereas it is in fact matter of behavior, i.e., some degradation will indeed occur, but at a rate that is unacceptably low. In drawing any methodological comparison, various direct chemical methods must be taken into account, and especially from the following points of view [41]-[45].

1. Energy costs and not only with regard to requisite temperature levels, but also the required duration of action and any energy expenditures associated with separation and enrichment steps, especially in the case of dilute wastewater.

2. Irregular waste from and how to cope with peak volumes or concentration. Microbial processes function best when volumes and concentrations are relatively constant. Large fluctuations, especially toward higher values, can often be managed with the aid of subsequent or parallel chemical process steps.

3. Mixed wastewater and the separation of partial (split) streams containing slowly degradable of interfering wastewater components [41],[42]. Major grounds for the separation and individualized treatment of specific production wastewater streams include the following:

   a. A certain partial stream is characterized by high volume and the presence of contaminant amenable to degradation by means of the simple and inexpensive low pressure wet oxidation process, resulting in a secondary wastewater suitable for introduction into a control biochemical wastewater-treatment plant.

   b. A partial stream contains wastewater components that degrade only slowly, suggesting that high-pressure wet oxidation. This process is quite elaborate and involves relatively high operating costs, so a more economical alternative might be to pursue chemical degradation only up to the point of suitable fragmentation, assigning the responsibility for further degradation, as above, to a central biochemical wastewater-treatment plant.

   c. A partial stream contains such a high level of salts that a cell culture would no longer be capable of functioning. Wet oxidation on the other hand might not be feasible either if in the case of certain salt-like companion substances the threat of corrosion rules out both low-and high-pressure oxidation. Such a case could be dealt with only by strictly thermal oxidation based on combustion, after appropriate concentration.

Generally, speaking, chemical treatment is restricted to special wastewaters characterized by components that are degraded too slowly in conventional waste-treatment plants or that interfere with the biochemical degradation of other substances. Chemical waste degradation processes listed in order of increasing operating temperature and pressure.

1. Atmosphere pressure wet oxidation by means of hydrogen peroxide, ozone, or air, with iron oxide or titanium dioxide as catalyst.
2. Low-pressure wet oxidation with air, based on an iron/quinone catalyst.
3. High-pressure wet oxidation with air, using copper as the catalyst.
4. Thermal oxidation, i.e., evaporation of water and combustion of the residue.

There is no such thing as a single best process, one clearly characterized by an ecological maximum and economic minimum. What is instead required is extensive experimental investigation to establish the limits and possibilities associated with each individual case, thereby providing a sound basis for meaningful comparisons. The question of how one should proceed with a particular wastewater source can then be answered reliably.

Thermal processes must generally ruled out in cases involving low concentrations of degradable substances because of excessively high specific energy costs per unit volume. The use hydrogen peroxide is often advantageous here, especially if the substances in question accumulate on an irregular basis. With higher contaminant concentrations, either high-pressure or low-pressure wet oxidation can be implemented. Evaporation followed by combustion becomes the method of choice if there is the added complication of high concentrations of inorganic salts. Wastewater of this type is derived largely from the chemical and materials industries.

The direct oxidation of organic compounds by hydrogen peroxide under acidic conditions is a well-known but relatively seldom-employed process. The oxidation potential of H_2O_2 can be increased above that of ozone through catalysis, usually with Fe^{2+}. Oxidation with hydrogen peroxide was investigated in the second half of the 20th century in conjunction with the treatment of municipal sewage[42] and waste water from industrial production, especially effluents containing sulfur and phenols [10], [43]. Successful hydrogen peroxide treatment of wastewater from hardening plants and tanneries has also been described.

In special advantage of this process is that the technological effort required is small, an especially positive factor in the case of dilute wastewater (<10 g COD /L). Non biochemical treatment is of special importance for wastewater fractions with low concentrations of contaminants, because the corresponding large volumes determine the size of the required treatment facility, and thermal processes are excluded because of the enormity of the associated energy requirement.

Oxidation with H_2O_2 proceeds at atmospheric pressure and room temperature within 60-90 min. Hydrogen peroxide is not toxic, is easy to handle, and decomposes into the environmental products oxygen and water. Large scale use of this otherwise virtually universal oxidizing agent is restricted by the high price of H_2O_2 itself, together with the fact that although self–decomposition proceeds with the formation of oxygen, the resulting oxygen contribution almost nothing to the oxidation process.

Hydrogen peroxide is not merely an oxygen-transfer agent that facilitates work in a homogenous aqueous phase, indeed, one should make every effort to avoid conditions favoring the self-decomposition of hydrogen peroxide:

$$H_2O_2 \Rightarrow H_2O + \frac{1}{2}O_2 \qquad (3)$$

More the point is the fact that catalysis by iron (molar ratio $H_2O_2 : Fe^{3+} = 15 : 1$, pH= 3.0), leads-as in the reaction of the Fenton reagent-to the formation of the H-O' radicals, and it is these that constitute the actual oxidizing agent.

The redox potential E_0 for the reaction sequence:

$$H_2O_2 \Rightarrow HO^{*} + OH^{-} \qquad (4)$$

$$Fe^{3+} + HO^{*} \Rightarrow Fe^{3+} + OH^{*} \qquad (5)$$

$$OH^{*} + H^{+} + e \Rightarrow H_2O \qquad (6)$$

has been determined to be 2.28 V. For the analogous reaction with ozone the redox potential is 0.21 V lower:

$$O_3 + 2H^{+} + 2e^{-} \Rightarrow H_2O + O_2 (E_0 = 2.07V) \qquad (7)$$

In the case of substance that oxidize only slowly even under these conditions, or via oxidation chains starting from them, it has been observed that self-decomposition of H_2O_2 - which is of no value for COD degradation-becomes appreciable. It is therefore necessary to establish the extent of utilization of introduced H_2O_2, and to complete this with the applied dosage in that particular case. The dosage should then be decreased until a utilization of nearly 100% is achieved [44]. If a dosage is calculated based on the stoichiometry corresponding to complete oxidation all the way to CO_2, an almost equivalent COD degradation can usually be achieved with 60% of the calculated dose, resulting in almost 100% utilization of the added H_2O_2.

The heat of reaction has also been determined under these conditions, leading to the following values established with a flow calorimeter, isothermal [45]:

$$\begin{aligned} Q_1 &= 4.59 cal / mgCOD \cong 19.22 J / mgCOD (\pm 3\%) \\ Q_2 &= 2.48 cal / mgH_2O_2 \cong 10.39 J / mgH_2O_2 (\pm 4\%) \end{aligned} \tag{8}$$

The extent to which different classes of substances differ in their behavior is illustrated by the data inTable 3. From an economic stand point, the selected conditions in this study were appropriate only for the last two substances listed. The process can be regarded as non-specific with respect to the various classes of organic substances typically encountered in wastewater. It also shows little dependence on the salt load.

The degradation of more than 100 individual compounds as well as authentic industrial wastewater has been described [46]. As a results of adsorption on flocculated material during the neutralization step the process leads to an effective reduction of COD roughly 10-15% greater than the level of true degradation.

The process can be operated continuously in the form of a cascade. On a pilot-plant scale (≤ 500L) the residence time is 60-90 min with a dosage of 20-80% H_2O_2/COD. Partial oxidation with maximum utilization of H_2O_2 is pursued only to the point at which the wastewater can be subjected to subsequent degradation in a biological wastewater-treatment plant without impairing biochemical processes there. The type and level of salts present is of some importance, even though the actual oxidation with H_2O_2 is largely insensitive to salts. Considering only the COD value of the wastewater, partial oxidation with H_2O_2 is still technically feasible with dilute effluent in the concentration range 0.5-10g COD/L, where thermal processes must be excluded because of the associated energy demand.

Radical formation based on hydrogen peroxide has also been achieved by UV irradiation for wastewater-treatment purposes [47].

Use ozone for the oxidative elimination of wastewater components has also been known for a long time. Ozone was used even earlier for the sterilization of air and the treatment of drinking water. Its effectiveness is highly dependent on the pH and is based essentially on two mechanisms.

1) So-called direct oxidation occurs under acidic conditions. This is fairly slow process, but the conversion can be accelerated greatly if the energy necessary for radical formation is provided in the form of UV high from a Hg low pressure radiation source.

$$\begin{aligned} &O_3 \Rightarrow^{h\nu} O^* + O_2 \\ &O^* + H_2O \Rightarrow^{h\nu} H_2O_2 \\ &H_2O_2 \Rightarrow^{h\nu} 2HO^* \end{aligned} \tag{9}$$

2) Alkaline oxidation also takes place via the intermediate formation of hydroxyl radicals:

$$\begin{aligned} &O_3 + HO^- \Rightarrow HO_2^- + O_2 \\ &HO_2^- + O_3 \Rightarrow^* O_2^- + O_2 + HO^* \\ &H_2O + O_3 +^* O_2 \Rightarrow HO^- + 2O_2 + HO^* \end{aligned} \tag{10}$$

Ozonization of organic halogen compounds leads to both inorganic halides, from dehalogenation at C-Cl, and the ensuing formation of new organic halides.

The formation of organic halides has also been detected in wastewater containing a combination of chlorine free organic materials and inorganic halides originating from the accompanying salt load. Compounds formed in this way generally reflect the presence of oxidizing agents, especially ozone or hydrogen peroxide.

In this system, oxidative degradation of C-H and C-C bonds and the formation of the new C-Halogen bonds in the presence of the appropriate salts is strongly pH dependent. Radical reactions produce the intermediate hypochlorite, which in alkaline medium also acts as an oxidizing agents. Chlorination at carbon is clearly apparent under acidic conditions, as implied by an equilibrium shift in the system

$$Cl_2 + H_2O \Leftrightarrow HOCl + HCl \tag{11}$$

This effect is even more pronounced in the case of the higher halides as shown by the following Data on C-Halogen formation in conjunction with oxidation (Table3).

TABLE 3. Halogen formation vs. treshold pH.

Halide	Threshold pH
Chloride	<5
Bromide	<10
Iodide	<15

Oxidative cleavage predominates in all three cases at a pH>4.

The oxidative treatment of wastewater containing salts requires the prevention of simultaneous and persistent halogenation. A high chloride content must be anticipated in seepage water from landfills and in certain effluent streams derived from product ion lines in chemical processing plants.

Bromide in drinking water leads to the formation of organobromine compounds upon ozonization. Special measures must be instituted in dealing with bromide- and iodide-containing wastewater derived from the processing of color negative films and the production of printing plates. In the case of iodide containing wastewater it is useful to remove the iodide prior to oxidation of other constituents. With bromide–containing wastewater it is possible to degrade the adsorabable organic halide (AOX) formed in situ during the course of oxidation by introducing an excess of oxidizing agent.

Radicals derived from natural ozone, and from natural, extracellular sources of hydrogen peroxide in the soil, especially the hydroxyl radical H-O* [48], also produce halogenated intermediates in the presence of ubiquitous salts, including halogenated metabolites. Some of the latter are intermediate products with a limited lifetime in the mineralization process, whereas others are persistent excretion products in the form of humid acids with halogenated aromatic groups. Similar processes are observed in the ocean and in inland waters.

Corresponding process in industry, i.e., oxidative degradation reactions with the simultaneous formation of intermediate organic halogen derivatives, has so far not been reported, although oxidative degradations in the formed natural decomposition and decay induced by natural oxygen sources together with biochemical catalysts: oxidizes, peroxidases have been simulated in the laboratory [49]. Specific proposals were developed in conjunction with this study for the industrially very important degradation of lignin in the wastewater released from paper and pulp mills.

Semiconductors catalysts such as the oxides of titanium, iron, or zinc, together with atmospheric oxygen as the oxidizing agent and sunlight as a source of the requisite activation energy, constitute systems for the treatment of dilute wastewater. These systems have already been investigated in the context photochemical degradation of organic substances in the atmosphere [48],[49].

Apart from studies involving individual substances, the behavior of actual wastewaters has also been investigated in a thin film fixed-bed reactor (Fig.6), including groundwater, seepage water, influents to biochemical wastewater-treatment plants, and the corresponding effluents.

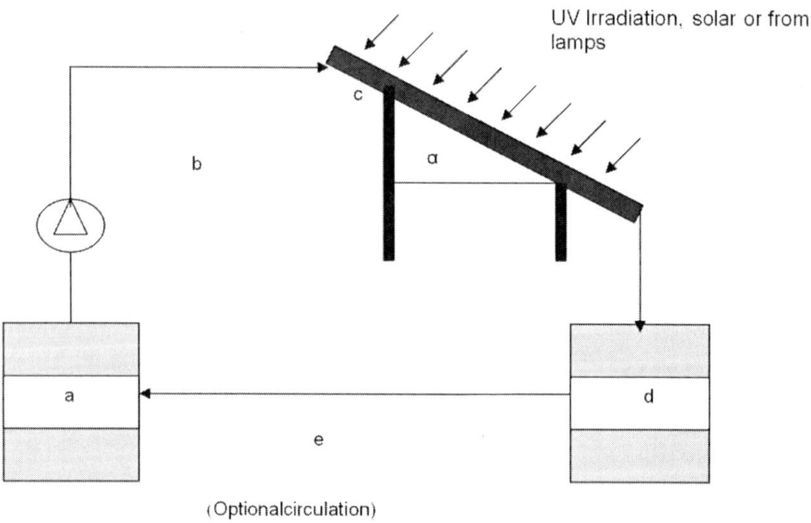

FIGURE 6. A thin-film fixed- bed reactor for solar detoxification.

A consideration of the associated energy factors shows that titanium dioxide absorbs light only at wavelengths below 390 nm because of its relatively large band gap of 3.2 eV. Accessible sunlight with a radiant power of 20-30 W/m^2 in the wavelength range 300-400 nm makes available 0.2-

0.3 photons per m^2 per h. Quantum yields of 0.1-10% have been reported at TiO_2 surfaces. Assuming an average quantum yield of 1%, this energy source would suffice for the degradation of 2-3 mmol of wastewater compositions per square meter of surface area per hour. For substances with molecular masses of ca.100 present in concentrations of ca. 1mg/kg, the potential wastewater throughput would be 200-300 L/h for an absorption surface area of one square meter. Lamp panels equipped with standard fluorescent lamps of the type developed for tanning benches have also been used as sources of near-UV radiation. The corresponding degradation reactions proceed more rapidly with hydrogen peroxide as a sole or supplementary source of oxygen.

Variants such as parabolic channel reactors have been found to be much less effective than thin-film fixed-bed reactors. Such chemical variants as hydrogen peroxide in conjunction with mixed oxides based on Fe (III)/Ti(IV) have led to better utilization of the available visible light, because α-Fe_2O_3absorbs at wavelengths under 540 nm.

Photo-reactivity has been found to increase in the presence of mixed oxides of this type as a result not only of photo-oxidation but also photo-addition, which may prove to be of some importance in the context of degradation intermediates.

Wastewater for which biological, chemical and physical purification processes either fail or prove too expensive can, if necessary, be purified by combustion of their organic constituents. In principle, all organic substances can be oxidized in a flame, and volatile inorganic substances formed can be separated in a flue-gas scrubber. Oxidation in a flame and the associated breakdown of organic materials depends on temperature, residence time, oxygen content, and turbulence in the combustion chamber.

The incineration of organic constituents is a specialized process for treating those industrial wastewater in which the organic constituents, and under some circumstances the inorganic constituents as well, are chemically utilizable at high temperature with the aid of atmospheric oxygen, the accompanying water matrix "gas phase oxidation" is also used in conjunction with this type of combustion to distinguish it from processes such as wet oxidation.

5. WASTE WATER MANAGEMENT SYSTEM

Waste water management involves analysis of elements, their attributes, behavior and parameters estimation, real time optimization and safety. Optimization provides optimal working conditions, services, troubleshooting, advanced control and hazard minimization. However, these will support people decision to prevent abnormal situation, not replace the people.

Waste water treatment operation makes history data base of manipulates entity variables. Supervision can make different service databases model. A model manager, which shows how does seek out a new way to create optimal waste water policy and how does model pretreatment and post treatment waste water network life cycle, and how does make management history was developed. Management systems will be available to analyze environment requirements and performed information processing in the aim objective achieving [51]-[58].

There is no safety reason to discontinue operation, when monitoring devices respond at the boundary between normal operating and admissible error ranges of waste water variables. Damage minimizing systems come into action when the waste water system is in no specified operation and when an undesired event occurs.

To use models to support decision making is proliferating in both the public and private sectors was the aim of this chapter. As powerful decision aids models can be both beneficial and harmful. At present, few safeguards exist to prevent model builders or users from deliberately carelessly data, or recklessly manipulating data to further their own ends. Perhaps more importantly few people understand or appreciate, the harm can be caused when builders or users, fail to recognize the values and assumptions on which a model is based or fail to take into account all the groups who would be affected by a model's results.

5.1. WASTE WATER SAFETY

Waste water from cities, industrial and the others manufactures are transported with various sewerage systems. Under infrastructured system understand process systems which using for recieving, collecting, evacuation and waste water treatment. Dependent of that how is collected and evacuation of waste water system can be general, separation and partical separated type.

At the general system all waste water is transported with one channel. The separation system are performed transport of all waste water from industry hausholders, and atmosphera separated channels. The partial separation systems mixing industrial water with hausholderswater or industrial and atmospheric water. Behinde these systems can made various combination in different parts of city. The waste water transport systems are consist from more connected elements. There are collectors, pipelines, valves, pumps which distributed in more streams as shown in Fig.7.

Maintenance waste water treatment and evacuation are very important. Safety of the waste water transport system is ecological significant. Because a diagnostic system for risk analysis and supervision of the waste water system is developed. In the safety analysis and operation the simulation getting started with data of the process components. For accident detection the derived model is forecasted the future behavior of the system and risk parameters are determined. The system is consisted of streams and process units data as well as the basic faults and symptoms.

This chapter illustrate the waste water transport safety protection system as shown in Fig. 8. In simulation, both qualitative and quantitative analysis are often applied together. Usually, qualitative decision efficiently made with symbolic and graphic information, and quantitative analysis is more conveniently performed by numerical information.

The diagnostic expert systems have important role how in risk analysis and accidents prevention of the production systems such as in transport systems (Fig.8). A diagnostic systems for supervision and maintaining process systems were developed [20]-[25].

Maintenance waste water treatment and evacuation are very important. Safety of the waste water transport system is ecological significant. Because a diagnostic system for risk analysis and supervision of the waste water system is developed. In the safety analysis and operation the simulation getting started with data of the process components. For accident detection the derived model is forecasted the future behavior of the system and risk parameters are determined.

The system is consisted of streams and process units data as well as the basic faults and symptoms. The diagnostic expert systems have important role how in risk analysis and accidents prevention of the production systems such as in transport systems. A diagnostic systems for supervision and maintaining waste water sewerage systems was developed in literature [1].

Operation makes history database of manipulates and object variables, symptoms and scenarios. Likely scenarios are typically generated by instantiating parameter values in a parametric model according to the given situation. Modelling of the risk parameters were involved uncertain processing.

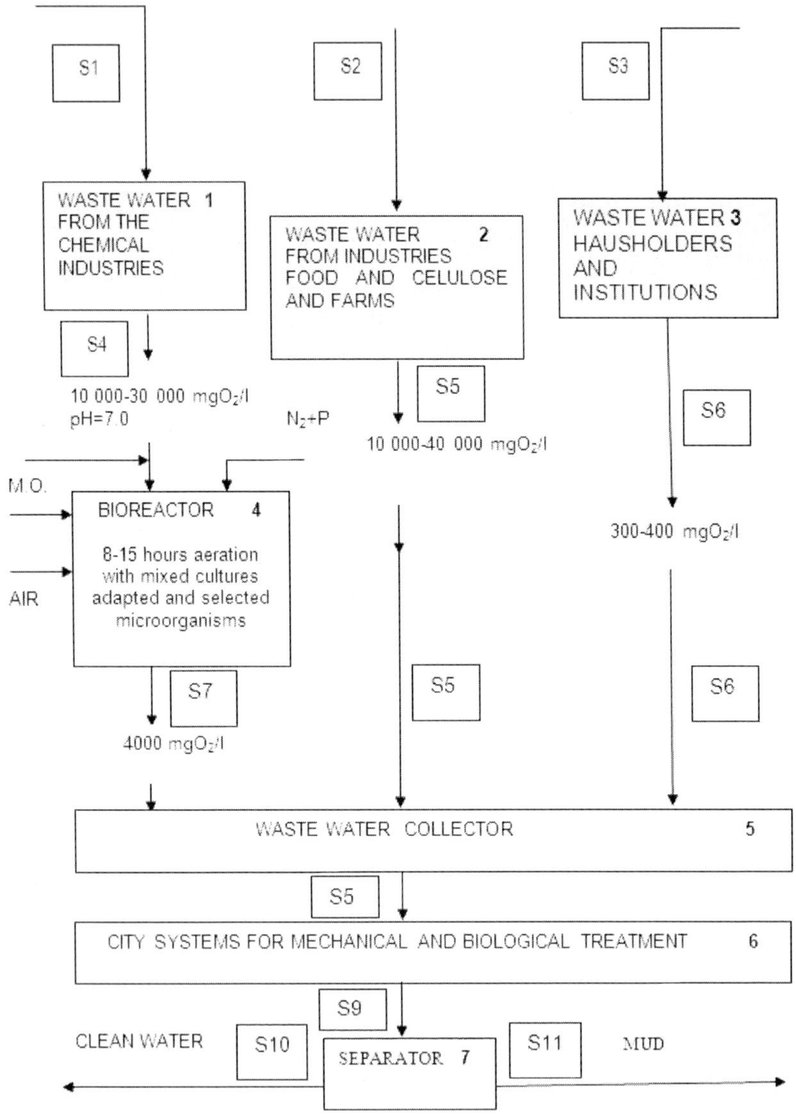

FIGURE 7. Waste water system evacuation.

The level of aggregation is defined by the modular component interconnections which define propagation paths of attributes within the system. Initial research starting by the phase of the development of a conceptual framework which will facilitate the modular specification of models, and second phase the development of a logic framework which will permit object using attributes and simulation techniques to be linked into executable models. The fault event of a system are in the first instance generally formulated in an IF-THEN form. This can be immediately reformulated using the operators AND, OR and NOT in Boolean form, if one can assume that the primary events have only two states existence and non-existence.

System identification involves identification of variables, element and equipment as well as material supplies. System variables are defined in three discret states low, medium and high. The equipment states are defined as blockage and leakage, system state as normal and does not work. Supply exists or not exists. As system variables consider pressure, flow and level.

The considered system consists of 11 waste water streams, four supply streams and 7 process units. A diagnostic system as a support decision system was built. The data base has involved data streams and process units data as well as the basic faults and symptoms which connecting by semantic network (Fig.8).

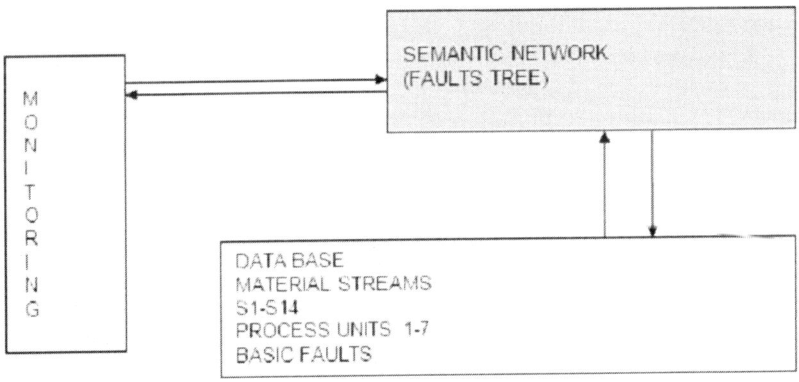

FIGURE 8. Sucture of the waste water diagnostic system.

The diagnostic system can assist to maintain waste water system safety transport. Waste water evacuation and treatment system safety is ecological very important problem. Fig.8 presents safety information support system, for waste water transport system diagnosis. The problem at hand is a problem of diagnosis, in which a major part of the solution consists of informing supervisor and action.

Such as systems can be used as a plant maintenance aids.

6. PROACTIVE DECISION OF THE WASTE WATER MANAGEMENT

A decision system on the process system which consists of operation and goal parameters was built. In the world mostly systems are used for solving structural problems, and they use retrospective view. These systems are used data from data base and directed to goal, but they have not view in front of and ability for unstructured problems solving. The process system for wastewater transport is unstructured. Goals and assumptions and decision support tools are examined such as events tree.

In recent years the study of systems management has successfully elucidated some basic techniques for generalizing concrete examples to more abstract descriptions [51]-[58]. These include heuristics for generalizing particular data types, candidate elimination algorithms, methods for generating decision trees and rule sets, back propagation of constraints through an explanation tree, function induction, and synthesis of procedures from execution traces [56],[57]. No coherent methodology has emerged for describing and categorizing management techniques to make them readily accessible to potential users. In fact there almost as many paradigms for management as there are systems and the variety of different, evocative, connotation laden words used to describe simple mechanism is one of the biggest problems in mastering the field. The apparent richness and variety of these approaches may give a misleading impression of a field teeming with fruitful techniques, with a selection of well defined methods for tacking and given problem [2]-[4].

From standpoint of knowledge engineering, concept management systems differ in their representation of concepts and examples, ways of biasing the search involved in determining a concept and model. This potential is rarely realized in practice because the field management is in turmoil and few useful general principles have been articulated. A model for decision making generates effective procedural, or rule based from a goal based architecture which further supports the development of secondary goals, as mental models.

From standpoint of knowledge engineering, concept management systems differ in their representation of concepts and examples, ways of biasing the search involved in determining a concept and model [58]. This potential is rarely realized in practice because the field management is in turmoil and few useful general principles have been articulated.

In this paper a model for decision making generates effective procedural, or rule based from a goal based architecture which further supports the development of secondary goals, as mental models.

The most important distinction in a knowledge acquisition framework is how systems represent what they manage. Knowledge representation has always been a central topic in process management [20],[21]. For both inductive and deductive concept the formalism above suggests representing concepts in an appropriately powerful form of logic (Fig 9). However, although formal logic provides a sufficient basis for deduction, as a foundation for induction it is at once too narrow and too powerful. On the other hand, logic is too powerful because the need to acquire knowledge automatically from environment and integrate it with what is already knows means that only the simplest representations are used by programs for system management. Any one representation will not encompass the broad application of concept management techniques, procedures or expressions composed of functions. These reflect fundamental formulations of computing that have been realized in logic, functional and imperative programming styles. Although equivalent in expressive power, the different representations are more or less appropriate for particular concept management problems, depending on the nature of the examples, background knowledge, the way the complexity of concepts are measured, and the style of interaction with the environment. For example, decision three are naturally represented as logic expressions, polynomials as functions and tasks as procedures. Functional representations incorporate the powerful mathematics available for numbers. Procedures embody the notions of sequencing, side effects and determinism normally required in sequential, real word tasks. There is an obvious overlap between logical and non-numerical function representations. For example, the concept of appending lists can equally well be written in logical and functional styles. The difference is that the logical form expression a pure relation without distinguishing input and output, while the functional representation acts on the input list to construct the output. Many management methods apply to examples that can be expressed as vectors of attributes in a form equivalent to propositional calculus. The values an attribute can assume may be nominal, linear, or tree structured. A nominal attribute is one whose values form a set with no further structure for examples the set of primary colors. A linear attribute is one whose values are totally ordered for example natural numbers. Ranges of values may be employed in descriptions. A tree structured attribute is one whose values are ordered hierarchically. Only values associated with leaf nodes are observable in actual examples: concept descriptions, however, can employ internal node names where necessary. Attribute vectors, propositional calculus are not powerful enough to describe situations where each example comprises a scene containing several objects.

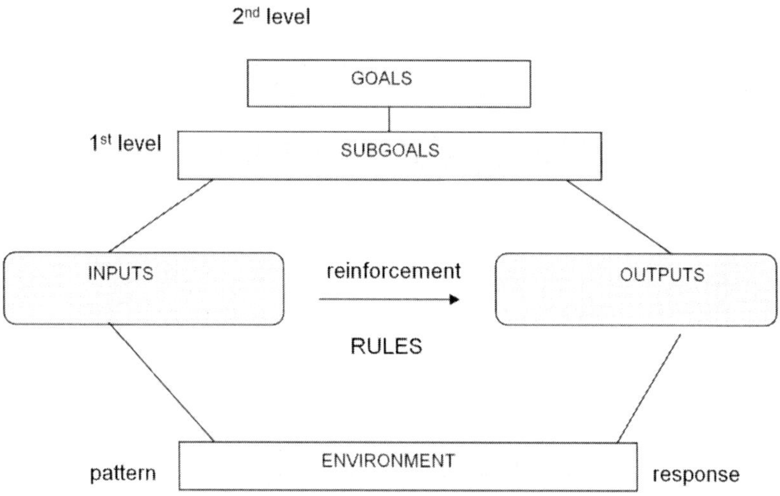

FIGURE 9. The goal based information system development.

Objects are characterized by their attributes. Moreover, pair wise relations may exist between them. This means that variables must be introduced to stand for objects in various relations. Such relations can be described by predicates which, like attributes, may be normal linear or tree structured. Objects and concepts are characterized by combinations of predicates.

Functional expressions include many natural laws, as well as relationships between quantities and parameters. Functional representations are appropriate for nested and recursive numeric or non-numeric expressions. Any functional relationship f(x) can be represented in logic and this is no surprise since the two forms are expressively equivalent. But in a framework for induction, it is preferable to treat functional expressions separately and omit explicit quantifiers. An important difference is that functional representations of concepts must be single valued, while logical expressions do not need to be this greatly affects the search space involved.

A more suitable form of representation might be the functional calculus, or some incarnation of it in pure of it in functional programming languages. Work on programming language semantics, also partly based on functional calculus and often coupled with function programming languages may suggest appropriate forms of expressly prohibit aspects that distinguish functional from procedural representations in the framework, namely side effects and reliance in sequential execution. Work on programming language semantics, also partly based on functional calculus and often coupled with function programming languages may suggest appropriate forms of representation. Existing function induction systems are specially designed for particular domains with little attention to more general forms.

Typical concepts in procedures category include procedures for assembly welding, and standard office procedures. The procedural formalism suitable for representing sequential execution where side effects, such as variable assignment and real world outputs like movements, make it vital to execute the procedure in the correct order. To describe a procedural language formally, a binding environment model of execution is needed, instead of the simpler substitution model that suffices for pure functional representations must be deterministic to be useful procedural concepts.

Note the generalization of execution traces into a procedure is not reducible to an equivalent problem involving the generalization of non-sequential input/output pairs as such a reduction will lose information about sequential changes in state.

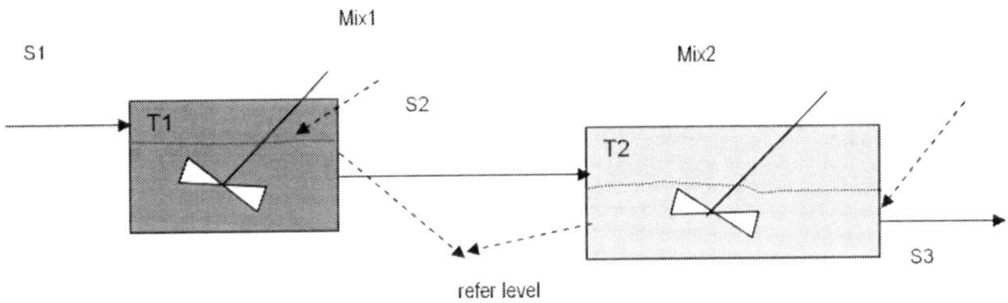

FIGURE 10. A simple wastewater flow system.

Let consider a simple process system wastewater transport as shown in Fig.10. The system consists of a two tanks, two mixers and pipes.

This system can be represented by qualitative events model expressed by logic algebra, M, B, and L are independent logic variables representing the basic events malfunction, blockage and leakage, respectively.

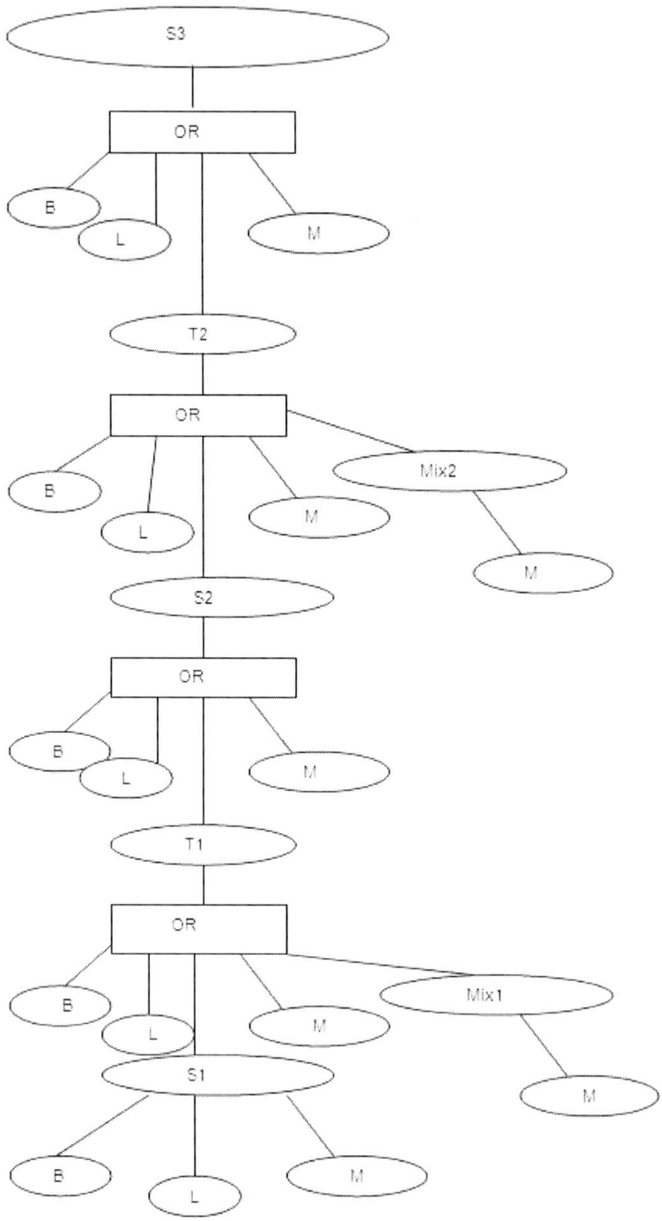

FIGURE 11. Fault tree diagnostic model.

The study of fault detection and supervision control of the flow system is concerned with designing a system that can assist a human operator in detecting and diagnosing faults as shown in Fig.11.

The two areas of model development and analysis are addressed through the discussion of generic simulation environment. The knowledge based simulation environment is an expression of some control law or cognitive theory. To the extent that the rule base is derived from set of assumptions about the environment and performance expectations, it is a belief system. However, in the existing form, the goals are not expressed and the underlying assumptions are not evident.

When expressed in hierarchical form the relationship that exist between goals and subgoals provide a basis for relating overall goal based system performance to specific assumptions about the variability and contribution of the supporting subgoals. In this form, the belief system is a full expression of some control theory in that the system's relationship with the environment, as expressed in a set of feasible state conditions, can be related either in overall system performance measures to be relationships and the subgoals that manage them.

Knowledge based system must represents information abstractly so that it can be stored and manipulated effectively, although experts have difficulty formulating their knowledge explicitly as rules and other abstractions. They find it easy to demonstrate their expertise in specific performance situations.

Concepts and examples are the output and input of the management acquisition system, what is manage and what is provided by a external agent. To be useful, a framework for representing concepts must provide knowledge engineers with methods for selecting appropriate representations for examples, concepts and background knowledge. Separate representations are required for examples and concepts.

Expert systems can be used to develop rules, based models and augment other types of models. Since can capture the experience, of experts, they can be used to forecast problems, advice, operators, validate data and ensure that the results from other are reasonable.

A functional approach to designing expert simulation systems was proposed many authors. They choose the differential games models is described using semantic networks. The model generation methodology is a blend of several problem solving paradigms, and the hierarchical dynamic goal system construction serve as the basis for model generation. Discrete event approach, based on the geometry of the games, can obtain the solution generally in much shorter time. Cooperation between systems is achieved through a goal hierarchy.

6.1. DECISION SUPPORT SYSTEM

The distinction between deductive and inductive concept management can be viewed as a modern reincarnation of the long philosophical tradition of distinguishing necessary from contingent truths.

The best way to solve complicated problem by expert systems is to distribute knowledge and to separate domain expertise. In such case, several expert systems may be used together. Each of them should be developed for solving a subdomain problem. Here, it is faced the problem of knowledge integration and data management (Fig.12).

Many expert systems can only be used alone for a particular purpose inflexibility. There are lack of coordination of symbolic reasoning and numeric computation, lack of integration of different expert system, lack of efficient management of intelligent systems and capability of dealing with conflict facts and events among the various tasks, being difficulty in modifying knowledge bases by end users other than the original developers.

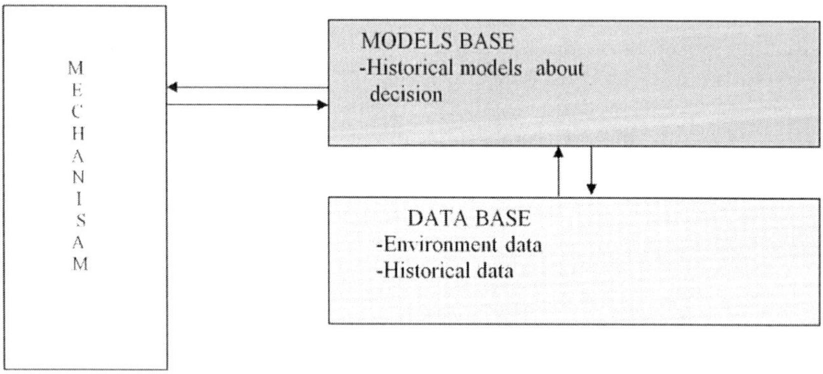

FIGURE 12. Decision support system building.

A knowledge based decision support system building is consisting from the following steps:

1. Wastewater system and treatment method identification
2. Goals and subgoals definition
3. Rules networking

4. Decision mechanism definition.
5. Monitoring system support

It can be indicated two possible approaches to complex contains modelling. The first, identified with structural knowledge, follows a deductive reasoning approach in which one tries to deduce from an existing theory model relationships for a given problem. The second, identified with a posterior empirical knowledge, follows an inductive approach in which one tries to develop a model from the sampled data. Ideally, these two approaches act as complementary stages of the modelling process.

One follows all six steps with model calibration and model validation serving as an empirical test bed for a prior model as a learning tool. Yet, in some situations characterized by difficulties in obtaining empirical data due to a budget and time constraints or peliminary scope of the analysis, the model specification may be reduced to the a priori stage.

In the traditional view of modelling the behavior of natural system these six steps, although logically connected, comprise separate tasks. Therefore, the development of an operational structure model is a rather lenghty and expensive undertaking. There is a strong need for a more integrated framework for modelling of system in general that would better link the steps leading to model development and implementation.

The best way to solve complicated problem by management systems is to distribute knowledge and to separate domain expertise. At this, several expert systems may be used together. Each expert system should be developed for solving a subdomain problem and it is faced the problem of knowledge integration and management.

The coordination of symbolic reasoning and numerical computation is required heavily for simulation with expert systems. A few developers tried to develop expert systems with conventional languages. Other suggested to field expert systems in conventional languages, in order to achieve integration. Another disadvantages is that the procedural language environment cannot provide many good features that the symbolic language provides, such as easy debugging allowance for interruption by human experts.

Many integrated intelligent systems are a large knowledge environment, which consists of several symbolic reasoning systems and numerical computation packages. They are under the control of a supervising intelligent system, namely, meta-system. The meta system manages the selection, operation and communication of these programs.

For example, decision support building for wastewater treatment management can outlines in four stages:

1. The physical control of information by computer, which became more complex as the volume of information increased, due to system growth, diversification and government regulation.

2. The management of automated technologies, where the introduction of data processing etc., led to fragmentation and uncoordinated activities.

3. Process system resources management, where data processing, office automation etc., converged along with central and personal computing resources.

4. Technology management, where the physical and technical management of information is integrated with decision making, planning and operations.

Waste water plant operation management support system aimed at helping engineers and managers optimize all phases of process plant design, operations, optimization and process safety. Decision support system is useful for supervision of process plant operations, real - time optimization, advanced interactive control and process hazard analysis. Some techniques are very important for implementing and evaluating decision support systems which expand such diverse areas as computer supported cooperative work, data base management, decision theory, economics, mathematical modeling, artificial intelligence, user interface management system and others [25].

Decision support system principles, concepts, theories and frameworks develops methods, tools, and techniques for developing the underlying functional aspects of a process plant management support systems, solver/model management in plant operation support systems, rule management and artificial intelligence in process plant system coordinating a plant management systems functionality within its user interface.

Decision of wastewater treatment process design and operation support system interfaces develops methods, tools, and techniques for developing the overt user interface, user knowledge, help of a facilities, coordinating interface event, with its functionality events.

Wastewater plant operation decision support system impacts shows economics, system measurements, decision support system impacts on individual users, multi participants users, evaluating and justifying.

A process plant management system was considered in papers [47]-[57]. Plant management decision support systems was studied in the paper [58].

Process plant information system and process safety management support system were investigated in the paper [5], [4],[17],[18].

The computer supported cooperative work with data base management and mathematical modeling and simulation of process plant provide intelligent plant management support. The model plant operation manager workflow is shown in Fig. 12. The plant manager workflow can be used for the new product quality improvement, the model operation network optimization and generating new data and specified tools. Real-time software need to build intelligent process management applications.

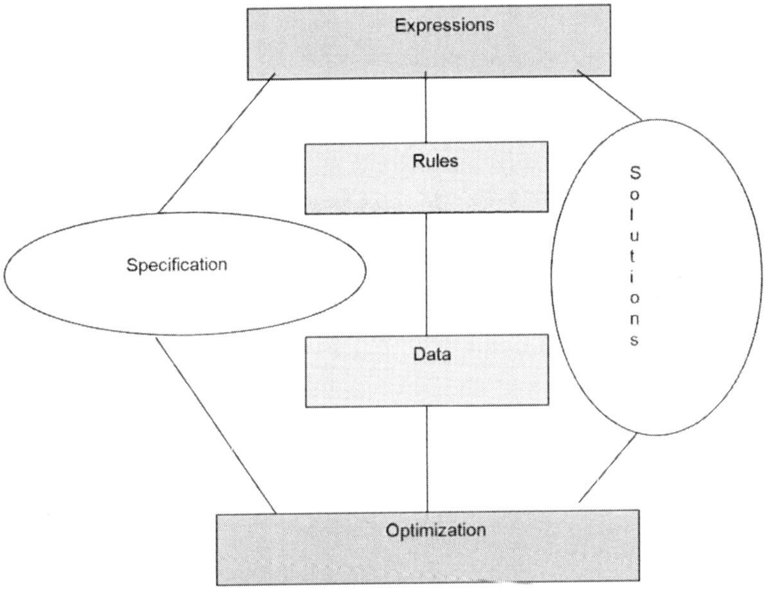

FIGURE 13. Management activities optimization.

Automatically create process operation models requires a windows based software tool for system identification which claimed to enable users to automatically create high fidelity models of physical systems, processes and plant.

System identification is technique that creates model of a plant operation process from input and output data, eliminating the need for detailed knowledge of the system physics. The new software automated the process, enabling engineers to perform modelling and simulation studies without having to create the underlying mathematical models from first principles.

Advanced commercial simulation systems also come, with in intelligent graphical user interfaces, which speed the development of error-free simulation problems and provide some help with thermodynamics and modelling.

Model operation manager shows how do you seek out a new way to create process operation, how do you model the plant life cycle and how do you make plant operation history in the process waste water treatment. A model plant manager is given by eq.(12).

$$MM = \sum_{i}^{n} M_i < T, P, A, F, E, Q(T), Q(P), Q(A) Q(F) Q(E) > \quad (12)$$

where *T* – set of elements, *P*-set of syntax rules, *A* –set of expression, *F*- set of

semantic rules, E - set stochastic events, *Q(J),J=T,P,A,B,E* changeable functionality, M_i- operation model and *MM*-management model.

In the process operation and control safety models "What if " help to build intelligent process management applications. They shoot stochastic events a day to day. It is improves plant operation reliability.

Process waste water operation involves sensitivity analysis of manipulate and object variables, parameters estimation, noises identification, dynamic simulation, detect process disturbance before they cause significant disruption (Fig.14).

Waste water treatment optimization provides optimal process condition, equipment services life, troubleshooting, advanced process control and minimization.

Superior customer support requests model evaluation. This future model is called "asses and control" as shown in Fig. 15. Where asses means:

- uncertainties not currently amenable to mathematical solution,
- people involvement, and
- advanced technologies support people

and control means:

- must work automatically,
- uncertainties can be handled rigorously,
- people eliminating, and advanced technologies support automatic operation.

Inherent softness exists in product pieces and forecast demands. In this frame it will throw lots of technology at these uncertainties or soft areas expert systems, neural networks, data reconciliation etc.. However, these will support people decision to prevent abnormal situation management, not replace the people.

Also, user can make a new overview this spreadsheet and printout. User can make input data tables, tables of consumption's and tables of flow rates.

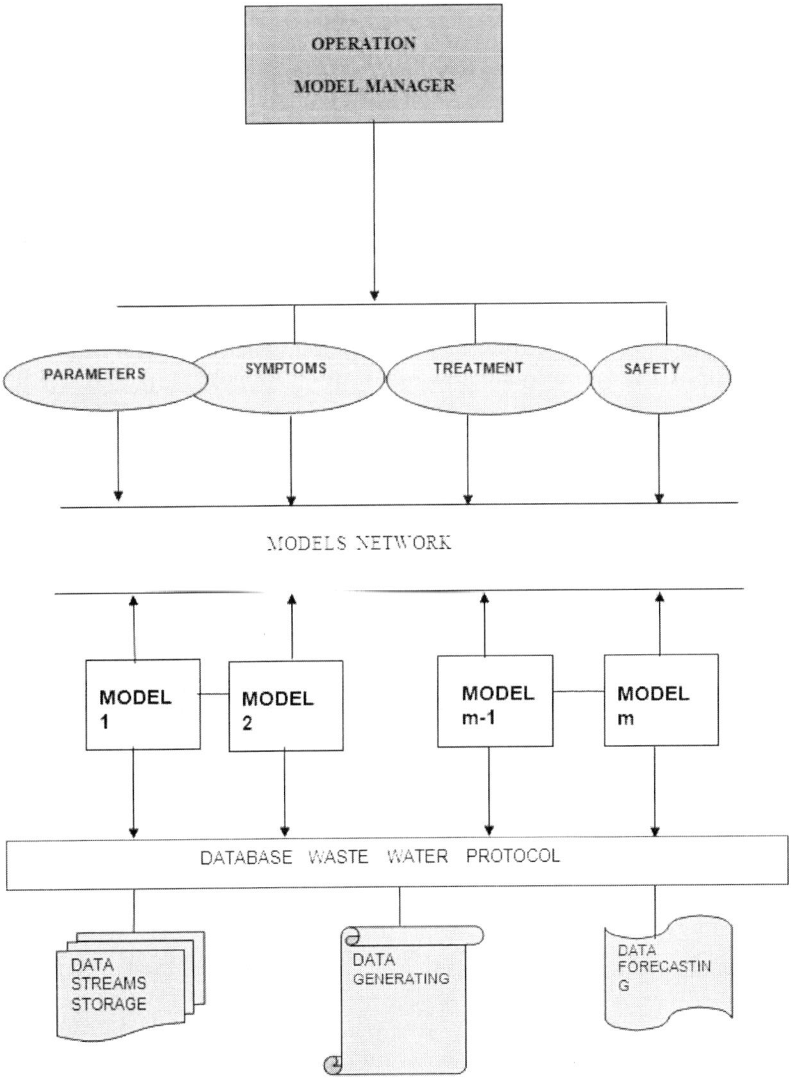

FIGURE 14. The wastewater operation model manager.

6.2. DATABASE MANAGEMENT

User can make different waste water plant database models. From the result summary user can extract and flow rates, conditions and consumption. If user whishes to see input and output streams results between stages should to mark on the appropriate block. When block menu appeared, select stream results for retrieval material and energy balances.

Databases protocol manages all databases, reports and tables. These tables could be distributed and extracted by multimedia databases. User can build hierarchical and relational process plant data bases for the plant management in concurrent operations (Fig.14).

The networks optimization requests maximum product capacity, minimum costs, concern the assignment problem, the matching and the minimum spanning trees, computer implementations and heuristics.

Process hazard analysis creates a resource allocation model by linking risk with cost and database of the values of basic events. In order to predict complete life cycle of the plant the plant life cycle reliability and the life cycle of the cost should be included.

Policy modeling emphasizes formal modeling techniques serving the purposes of decision making. These systems Computer aided process engineering-CAPE, Computer aided control-CAC, Computer integrated manufacturing-CIM, Computer aided safety-CAS make distributed computing.

The use of mathematical models to support decision making is proliferating in both the public and private sectors. Advances in computer technology and greater opportunities to learn the appropriate techniques are extending modeling capabilities to more and more people.

As powerful decision aids process operation models can be both beneficial and harmful. At present, few safeguards exist to prevent model builders or users from deliberately carelessly, or recklessly manipulating data to further their own ends. Perhaps more importantly few people understand or appreciate, the harm can be caused when builders or users, fail to recognize the values and assumptions on which a model is based or fail to take into account all the groups who would be affected by a model's results.

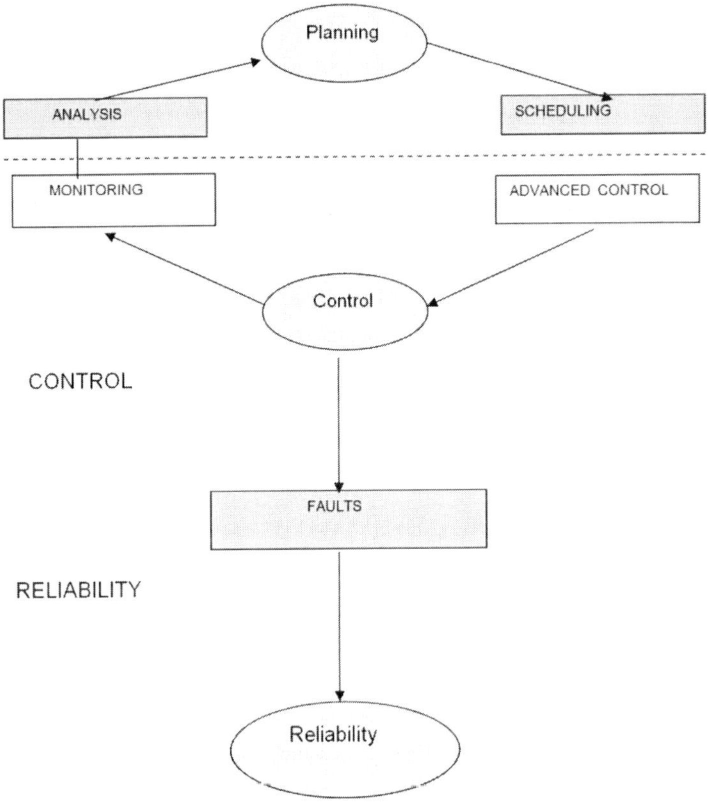

FIGURE 15. Asses, control and reliability.

Simulation manager models provide a setting for dialog and show the need to continue and define a vocabulary for exploring process operation. It will become increasingly important for model builders and users to have a clear and strong code to guide process plant operation.

The computer supported cooperative work provides explanation based process learning systems. As a cooperate function training is a subsystem within the plant's large organizational system. When training strategies ensure acquisition of knowledge, skills and attitudes which results in improved performance or safety on the plant operation, the training subsystem makes a positive contribution to organizational goal and effectiveness. Process performance, then, is the criterion of success in training.

7. NOTATION

A-set of expression

c-concentration, mg/L

F- set of semantic rules

k-specific constant

M-model

MM-model manager

P- set of syntax rules

Q- function

T- set of elements

X-equilibrium amount loading

8. INDEX

i - component

9. ABBERVATION

COD-contaminants oxidative degradation

AOX- adsorabable organic halide

REFERENCES

1. J. Savkovic-Stevanovic, 2009 Waste water transport system safety, Proceedings of the 2nd International Conference on Maritime and Naval Science and Engineering 71 76 Brasov, Romania
2. J. Savkovic-Stevanovic, L. Filipovic-Petrovic, R. Beric, 2011 Network service systems for chemical engineers International Journal of Mathematical Models and Methods in Applied Science 5 1 105 112

3. J. Savkovic-Stevanovic, 2010 Reliability and safety analysis of the process plant Petroleum and Coal 52 2 62 68 1337-7027

4. J. Savkovic-Stevanovic, 2009 Decision support system extraction, Proceedings of the 11th WSEAS International Conference on Mathematical and Computational Methods in Science and Engineering Cambridge, London 44 49

5. J. Savkovic-Stevanovic, 2009 Process Safety in the chemical processes 489 584 Chapter in the Book-Process plant operation, equipment, reliability and control, Editors,C.Nawoha, M. Holloway, and O. Onyewuenyi 768 pages John Wiley &Sons New York, USA 978-1-11802-264-1

6. C. J. Chicken, R. M. Hayns, 1989 The risk ranking technique in decision making Pergamon press

7. Ullmann's Encyclopedia of Industrial Chemistry 1995 B8 VCH Verlagsgesellschaft 3-52720-138-6

8. J. Savkovic-Stevanovic, T. Mosorinac, S. Krstic, 2006 Process risk analysis operation modelling Comput. Ecol. Eng 2 1 31 37 1452-0729

9. J. Savkovic Stevanovic, 2010 An advances learning and discovering system Transaction on Information Science and Applications 7 7 July 1005 1014 1790-0832

10. J. Savkovic-Stevanovic, S. Krstic, 2006 Risk reduction support system of the phenol recovery plant Petroleum and Coal 48 6 1337-7027

11. J. Savkovic-Stevanovic, Procedure for water depoluted from phenol WIPO patent, PCT,WO2006/096891

12. J. Savkovic-Stevanovic, Drinking water producing from see water WIPO patent, PCT,WO2007/115339

13. J. Savkovic-Stevanovic, 2008 Procedure cresol removal from water SRB, 2006D3/00 Bul.3 568

14. J. Savkovic-Stevanovic, 2004 The Chemical Plant Management Qualitative Support System Chem.Ind. 58 123 127

15. J. P. Black, 1967 Improvements behavior of aluminum species, in dilute dispersed kaolinite systems J.Am.Water Works Assoc 1173 1183

16. J. Savković-Stevanović, 2005 Network models for the plant management Comput.Ecol.Eng 1 87 93 1452-0729

17. J. Savkovic-Stevanovic, 2009 Knowledge base information processing, WMSCI2009The13th World Multi conference on Systemics, Cybernetics and Informatics Orlando, Florida, USA 10-13 July

18. J. Savkovic-Stevanovic, 2007 Informatics Faculty of Technology and Metallurgy, Belgrade University Belgrade

19. J. Savković-Stevanović, 2009 Decision support system extraction 11th WSEAS International Conference on Mathematical and Computational methods in Science and Engineering MACMSE09, ID 639 126 Morgan State University, USA November 7-9 Baltimore

20. J. Savković-Stevanović, T. Mošorinac, 2009 Management of the chemical industrial plant 1st International Congress on Engineering, Materials and Management in the Process Industries Jahorina, Serbian Republic, Bosnia and Hercegovina 4th October

21. S. Krstic, J. Prodanovic, J. Savkovic-Stevanovic, 1994 Expert added simulation in process operation Comput. Chem. Eng 18 767 771

22. N. Takama, T. Kuriyama, K. Shiroko, T. Umeda, 1980 Optimal water allocation in a petroleum refinery Comput. Chem. Eng 4 251 258

23. J. Djurovic, 2011 Decision operation of a process systems Comput. Ecol. Eng 7 2 49 56 1452-0729

24. J. A. Gougen, 1969 The logic of inverse concept Synthese 19 325 373

25. J. Savkovic-Stevanovic, 1999 Process engineering intelligent systems Chapter 3 RAJ Publishing Memphis,U.S.A

26. J. Savkovic-Stevanovic, 1987 Information Systems in the Process Techniques Scientific book Belgrade

27. M. A. Tubaisht, S. K. Madria, B. Bhargava, 2002 Pereformance evaluation of linear hash structure model in a nested transaction environment J. Systems and Software 63 219

28. T. Mošorinac, N. Boškovic-Vragolović, M. Milivojević, 2011 Expert systems in logistic simulation, Proceedings SRBISIM2011-the 2nd Conference on Modelling and Simulation 143 149 Belgrade, Serbia 28-29 September

29. J. Savkovic-Stevanovic, 2002a Process plant management system Proceedings of the IES2002-The 13th Symposium and Seminar on Infirmation and Expert Systems in the Process Industries 8 27 Belgrade, Serbia 30-31,Oktober

30. J. Savkovic-Stevanovic, 2000 Neuro-fuzzy system modeling and application for a distillation plant Chem. Ind. 54 389 392

31. J. Savkovic-Stevanovic, 2002 A neuro-fuzzy controller for product composition control of the ethanol recovery distillation plant CHISA-15th Congress of Chemical and Process Engineering 1102 Praha 25-29 Aug

32. J. Savkovic-Stevanovic, 2003 Plant Management Information Systems Proceedings of the IES2003-The 14th Symposium and Seminar on Information and Expert Systems in the Process Industries 32 36 Belgrade 23-24

33. J. Savkovic-Stevanovic, 2003 The Chemical Plant Management Support System The Industrial Simulation Conference ISC 2003, EUROSIS Valencia, Spain June 9 12

34. J. Savkovic-Stevanovic, 1990 Industrial plant management system Bull. Chem. Technol. Macedonia 22 159

35. L. Svarovsky, (ed.) Solid- liquid separation 3rd ed. Butterworths, London

36. W. W. Eckenfelder, 1989 Industrial Water Pollution Control 2nd ed. McGraw - Hill New York

37. Report of the Committe on Water Quality Criteria 1968 Federal Water Pollution Control Administration U.S.A. Department of the Interior April 185

38. Kirk-Othmer Encyclopedia of Chemical Technology 1970 John Wiley and Sons, Inc

39. B. G. Richard, C. P. Straub, 1974 Editors, Handbook of Environmental Control, Wastewater treatment and Disposal IV CRC Press Clevelend, Ohio

40. Nalco Chemical Company Ion Exchange Division, Chikago, Illinois, The Su-biSul Process Bulletin Z-18 TD Index 680.93

41. W. Simmler, 1984 Chem. Ing. Tech 56,117

42. Bahnemann 1994 Nachr. Chem. Tech. Lab 42 378

43. W. Simmler, 1982 Ecotoxicol. Environ. Saf. 6 399

44. W. G. Strunk, 1979 Hydrogen peroxide treats diverse wastewaters Ind. Waste 25 1 32 35

45. W. Simmler, 1984 Chemistry of the future Pergamon Press Oxford 63 80

46. E. Gilbert, 1975 Uber den Abbau von organischen Schadstoffen im Wasser durch Ozon Vom Wasser 43 275 290

47. W. W. Glaze, J. W. Kang, D. H. Chapin, 1987 The chemistry of water treatment processes involving ozone, hydrogen, Peroxide, and ultraviolet radiation Ozone, Sci. Eng. 9 335 352

48. J. Rudolph, 1993 Korres, Abwasser 40 1289

49. W. W. Glaze, J. F. Kennecke, J. L. Ferry, 1993 Environ. Sci.Technol. 27 177

50. Verband der Chemischen Industrie Verfah rensbericht, Abwasserbehandlung Frankfurt am Main March 1

51. J. Savkovic-Stevanovic, 2004 The process plant management risk reduction support system Proc. of the 11th European Symp. on Loss Prevention, Praha 223 May-3 June

52. J. Savkovic-Stevanovic, 2005 Process hazards analysis and safety standards Comput. Ecol. Eng 1 1 47 59 1452-0729

53. J. Savkovic-Stevanovic, 2005 Risk assessmment and reliability analysis Comput. Ecol. Eng 1 1 69 75 1452-0729

54. J. Savkovic-Stevanovic, 2005 Process safety management Comput. Ecol. Eng. 1 1 35 46 1452-0729

55. J. Savkovic-Stevanovic, 2005 Cognitive reliability analysis of the process plant Computer Aided Process Engineering-15, Ed. D. Puijagner, A. Espuna 22 t.p.1718 Elsevier Science 342 347

56. J. Savkovic-Stevanovic, 2007 Process plant risk analysis and modelling Computer Aided Process Engineering-17, Eds. V. Plesu and P.S. Agachi 24 Elsevier Science 1229 1233 978-0-44453-157-5 1570-7946

57. J. Savković-Stevanović, 2009 System safety analysis Proceedings of the IES2009-19th Symposium and Seminar on Information and Expert Systems in the Process Industries 1 32 30-31 October 978-8-69110-113-8

58. J. Savkovic-Stevanovic, 2011 Simulation in the knowledge base and data base network AIKED11International Conference on Artificial Intelligence, Knowledge, Engineering and Data bases ID650-650 Cambridge, London 20-22 Feb

5
CHAPTER

The Need for Agricultural Water Management in Sub-Saharan Africa

Kbrom Ambachew Gebrehiwot*, Mehari Gidey Gebrewahid

Department of Water Resources and Irrigation Engineering, Institute of Technology, Haramaya University, Haramaya, Ethiopia

ABSTRACT

This paper discusses the current and future conditions that affect water resources and the constraints of water (agricultural) management in sub-Saharan Africa and suggests remedial measures to be considered by policy makers. The pressure on the quantity and quality of water resources is rising in sub-Saharan Africa due to the increased demand of water for agriculture and other purposes as a result of increase in population and food demand. The availability of water is also under threat from changing climate and as a result, water scarcity is expected in many countries in sub-Saharan Africa. On the other hand, the availability of water for agriculture is expected to further shrink due to the increasing demand of water for other purposes like industry, manufacturing and environmental requirement. The current poor efficiency rate of irrigation systems and massive expansion of irrigated area is expected to further exacerbate the water scarcity. Hence, a water management policy focused on maximization of water use efficiency and water productivity should be prioritized in order to meet the food demand of the growing population and cope with water scarcity problems. Engineering and management intervention integrated with strong society awareness and participation is considered very crucial in enhancing water use efficiency and crop water productivity.

Keywords: Water Scarcity, Water Management, Water Use Efficiency, Water Productivity, Sub-Saharan Africa

1. INTRODUCTION

World population has risen steadily from a total of 2.5 billion in 1950, to 7.4 billion in 2015 [1] . 16% of this growth comes from sub-Saharan Africa as the region has grown from 180 million to 962 million people from 1950 to 2015. This is about 12 million people a year for the past 65 years or approximately 782 million people in 65 years. According to the 2015 UN Revision, the current world population of 7.4 billion is expected to increase by 2.3 billion people within the next thirty five years, reaching 9.7 billion in 2050, and to further increase to 11.2 billion by 2100. Not surprisingly, by 2050 and 2100, the population of sub-Saharan Africa could be as large as 2.1 and 3.9 billion. This means 77% of the world population growth from 2015 to 2100 will be from sub-Saharan Africa.

This population growth would inevitably pose additional demand for food [2] [3] . As shown inFigure 1 and Figure 2, majority of countries with higher population growth rate, for example, Ethiopia, Kenya, Uganda and Nigeria, are precisely those showing high level of undernourishment as a result of inadequate food consumption. Nonetheless, these countries are on the verge of development and the food consumption pattern is expected to increase in the near future. The high population growth rate coupled with increased per capita food consumption will highly increase the food demand of the region which cannot be fulfilled simply with the existing production level.

In sub-Saharan Africa, food consumption of cereals such as maize, millet, sorghum, and teff (in Ethiopia) is the mainstay of diets [4] . As presented in Table 1, the demand for cereals is higher than the production in many of the sub-Saharan Africa countries. The gap between production and consumption would be more pronounced if other food groups, like starchy roots, vegetables, meat, egg, and others, are considered. To fill this gap and ensure food security dramatic increase in agricultural production is inevitable.

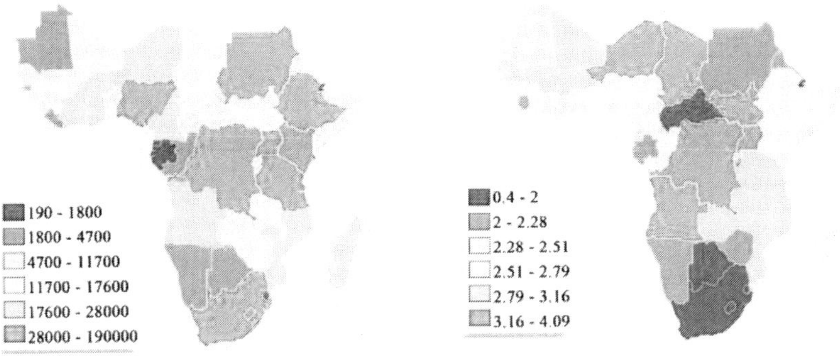

Figure 1. Total population (thousands) (left) and population growth rate (%) (right) in 2015 (source: United Nations) [1] .

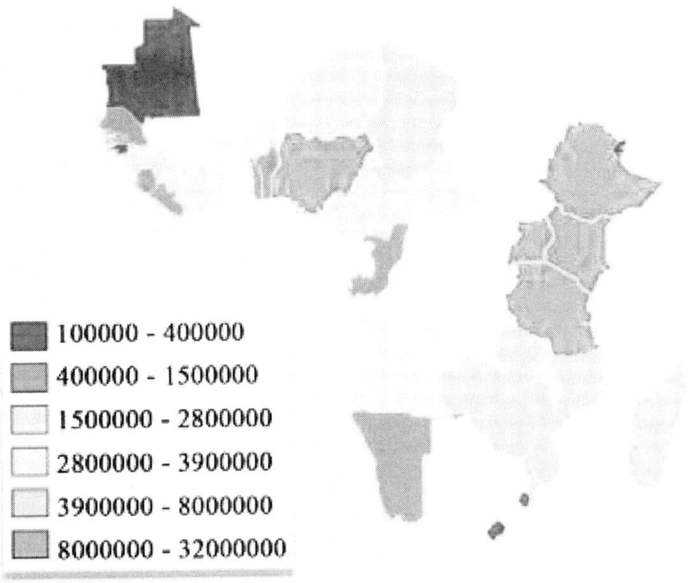

Figure 2. Number of people who are undernourished in 2015 (Source: World Bank [4]).

Table 1. Food (cereals) production and consumption in sub-Saharan Africa (Source: FAO food security data by food groups/items, June 2012).

Country	Production (kg/person/year)			Consumption (Kg/person/Year)			Deficit		
	1997	2002	2007	1997	2002	2007	1997	2002	2007
Angola	32	41	45	59.9	72.3	81.0	−27.9	−31.3	−36.0
Benin	130	136	130	108.0	104.0	116.1	22.0	32.0	13.9
Botswana	42	16	18	123.7	109.5	109.5	−81.7	−93.5	−91.5
Burkina Faso	215	233	242	223.7	219.7	233.6	−8.7	13.3	8.4
Burundi	43	37	35	30.3	28.8	35.8	12.7	8.2	−0.8
Cameroon	86	84	95	89.4	96.7	108.8	−3.4	−12.7	−13.8
CAR	35	45	53	41.6	48.9	57.3	−6.6	−3.9	−4.3
Chad	121	128	181	108.8	118.3	144.5	12.2	9.7	36.5
Congo	4	5	6	44.2	75.2	77.0	−40.2	−70.2	−71.0
Comoros	24	21	21	74.5	74.5	75.2	−50.5	−53.5	−54.2
DRC	30	27	23	34.7	32.9	39.8	−4.7	−5.9	−16.8
Djibouti	0	0	0	111.3	125.9	138.7	−111.3	−125.9	−138.7
Eritrea	31	31	74	127.8	120.8	134.0	−96.8	−89.8	−60.0
Ethiopia	145	132	168	123.0	138.3	142.7	22.0	−6.3	25.3
Gabon	25	20	23	94.2	101.8	109.5	−69.2	−81.8	−86.5
Gambia	85	120	117	139.8	142.7	147.5	−54.8	−22.7	−30.5
Ghana	95	87	81	82.5	87.6	97.5	12.5	−0.6	−16.5
Guinea	143	167	212	127.4	127.0	127.0	15.6	40.0	85.0
Kenya	101	93	100	120.1	117.5	120.5	−19.1	−24.5	−20.5
Lesotho	103	97	52	216.8	221.9	227.4	−113.8	−124.9	−175.4
Liberia	34	33	35	97.8	89.1	97.5	−63.8	−56.1	−62.5
Madagascar	137	121	145	111.7	117.9	129.6	25.3	3.1	15.4
Malawi	164	167	177	147.8	142.4	143.1	16.2	24.7	33.9
Mali	205	205	274	181.0	187.6	199.7	24.0	17.4	74.3
Mauritania	78	43	50	164.3	159.5	169.0	−86.3	−116.5	−119.0
Mauritius	0	0	1	152.2	155.1	161.0	−152.2	−155.1	−160.0
Mozambique	79	76	66	91.6	92.7	101.5	−12.6	−16.7	−35.5
Niger	207	247	281	200.0	201.1	197.5	7.0	45.9	83.5
Nigeria	185	156	181	136.9	133.6	145.6	48.1	22.4	35.4
Rwanda	31	33	39	33.9	32.1	43.4	−2.9	0.9	−4.4
Senegal	100	87	86	154.0	152.6	164.3	−54.0	−65.6	−78.3
Sierra Leone	78	54	136	107.3	113.2	112.1	−29.3	−59.2	23.9
Sudan	134	115	157	138.3	123.0	130.7	−4.3	−8.0	26.3
Swaziland	138	81	50	133.6	124.1	127.8	4.4	−43.1	−77.8
Tanzania	132	130	133	106.2	106.9	104.0	25.8	23.1	29.0
Togo	142	141	137	116.1	118.6	123.7	25.9	22.4	13.3
Uganda	80	88	84	59.5	58.8	63.5	20.5	29.2	20.5
Zambia	127	90	116	157.3	142.0	136.9	−30.3	−52.0	−20.9
Zimbabwe	192	141	120	148.6	144.5	157.0	43.4	−3.5	−37.0
South Africa	272	280	227	178.5	186.2	186.9	93.5	93.9	40.1
Namibia	68	59	64	130.7	132.1	131.4	−62.7	−73.1	−67.4
Côte d'Ivoire	70	62	58	96.0	89.4	93.4	−26.0	−27.4	−35.4
Guinea-Bissau	111	99	113	139.4	146.7	146.7	−28.4	−47.7	−33.7

Approximately 40% of the world's food comes from 17% of the world's cropland that is irrigated which consumes more than two-third of the world's developed water supplies [5] - [8] . The increase in population and the food demand as well as the changing consumption patterns toward a higher nutrition will be posing challenges to agricultural production [9] . A FAO analysis [10] of 93 developing countries expects agricultural production to increase over the period 1998-2030 by 49% in rain fed systems and by 81% in irrigated systems. According to World Bank [4] , agricultural production could nearly triple in sub-Saharan Africa by 2050. In this regard, arable land expansion will remain an important factor in the growth of crop production in the region. In developing countries, it is expected to observe a projected net increase in the arable area of some 107 million ha (from 968 in 2005/07 to 1075 in 2050) of which the 51 million (48%) is expected to take place in sub-Saharan Africa [4] .

In sub-Saharan Africa, the water use efficiency estimated for the year 2005/07 and projected for the year 2050 is 22 and 25 percent respectively, which is by far less than the world average which is 44 and 46 percent respectively [11] . Although, the current water withdrawal for agriculture and other purposes is lower than the total annual renewable water resource, this low water use efficiency will undoubtedly challenge the sustainable use and management of water in the region. Thus, expanding the irrigated area under the present agricultural water management characterized with low irrigation efficiency rates will result in severe water scarcity in the coming decades.

From Figure 3, it is clear that countries with larger total renewable water resources are also having a huge population, resulting in a lower total per capita water availability. In 2014, 16 SSA countries were already under water stress, water scarcity and absolute scarcity conditions according to the "Falkenmark indicator" or "water stress index". Based on the per capita usage, the water conditions in an area can be categorized as: no stress (> 1700 m^3 per capita), stress (1000 - 1700 m^3 per capita), scarcity (500 - 1000 m^3 per capita), and absolute scarcity (<500 m^3 per capita) [12] . As shown in Table 2, the per capita renewable water resource was declining with time due to the increase in population. As predicted by Falkenmark [13] , most of the East and some of the South African countries have already been water stressed. The projection by Falkenmark [13] indicates that eight of the East and some of the South African countries will arrived at absolute scarcity by 2025 and five of the West African countries will become at least water stressed in the early decades of the twenty first century.

Not surprisingly, the water resources of some countries had already been declined and changed from no stress to stress and scarcity levels. For instance, in 2007, Comoros, Nigeria and Uganda were under no stress condition. However, in 2012 and 2014, their water resource was declined and changed to the stress category. Similarly,

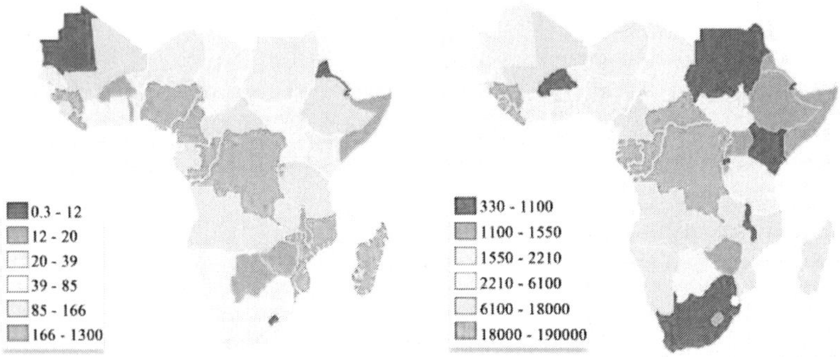

Figure 3. Total renewable water resources, BCM, (left) and total renewable water resources per capita, m³/person/yr, (right) in 2014 (Source: FAO Aquastat 2015).

Table 2. Per capita renewable water resources (Source: FAO Aquastat 2015).

Country	Water resources (m³/per/yr)		
	2007	2012	2014
Djibouti	375.50	348.80	339
Kenya	813.20	711.00	674
Burkina Faso	948.40	820.20	775
South Africa	1035.00	980.20	966
Malawi	1260.00	1086.00	1027
Sudan	No data	1016.00	975
Rwanda	1340.00	1161.00	1099
Eritrea	1404.00	1193.00	1119
Burundi	1506.00	1273.00	1196
Ethiopia	1517.00	1330.00	1264
Lesotho	1545.00	1473.00	1440
Zimbabwe	1570.00	1457.00	1370
Somalia	1650.00	1442.00	1360
Comoros	1896.00	1671.00	1596.00
Nigeria	1944.00	1695.00	1603.00
Uganda	1956.00	1654.00	1547.00

South Africa's water resources status was changed from stress category in 2007 to scarcity category in 2014. Nevertheless, the attention given to water management in sub-Saharan Africa is still inadequate. As a result, over exploitation of water resources by different sectors is shadowing the efforts made to ensure sustainable water use in the region. Considering the global figure, withdrawal of water by about a quarter of the irrigated systems is above the regeneration rate [14] . This coupled with the poor property rights specifications on water resources and inefficient irrigation practices is creating a potential for serious water shortages. Moreover, with the growing demand of water for industry, manufacturing and urban sectors, combined with environmental problems, water available for agriculture will decrease in the future [15] - [18] . On the other hand, as irrigation (agriculture) is vitally important in meeting the food and fiber needs for a rapidly expanding world population, the demand for agricultural water is continuously growing [3] [7] [19] . Additionally, the rapid growth in population and income requires large water investments and financing that result in further pressure on the quantity and quality of water resources [20] . The problem of providing the food for a much greater world population, therefore, becomes focused on the area of growing more on existing water and land resources [21] .

2. WHAT ARE THE AREAS OF IMPROVEMENT IN WATER MANAGEMENT?

Many sub-Saharan Africa countries must develop more than twice the amount of water they currently use to meet reasonable future requirements [21] . For instance, Ethiopia must double its cereal production by 2025 to meet the food needs of its rapidly growing population and secure food self sufficiency which in its turn escalate its agricultural water demand [22] . Considering the fact that 90% of staple food for sub Saharan Africa comes from rain-fed farming systems, [23] cited in [24] recommended that much greater emphasis will have to be given to increasing the productivity of global rain-fed agriculture. However, rain-fed agriculture remains risky because of spatial and temporal variability in rainfall, water scarcity, droughts, soil erosion by wind and water, low investment, or high population pressure [25] . Moreover, there is a significant difference in crop yields between irrigated and rain-fed agriculture. In developing countries, grain yields from rain-fed agriculture are 1.5 t/ha compared to 3.1 t/ha from irrigated agriculture on average [26] cited in [9] . Hence, due attention will have to be given to better management of water resources in both irrigated and rain-fed agriculture. In this regard, instead of trying to supply the growing demand for water with new sources, improvement in the water-use efficiency should be the main focus in water-management policies [18] [27] . To increase water-use efficiency, several practices can be adopted. Water-saving techniques, advances in irrigation technology (e.g., drip irrigation) and management, leaving crop residues on the soil surface and planting cover crops and adequate tillage practices are considered as potential measures to improve water use efficiency and adapt to climate change impacts [9] [28] - [30] . However, with the limited and shrinking water availability, the increasing water requirement, due to expansion of irrigable lands as a result of growing population and food demand, cannot be reversed by improving the water use efficiency only. Hence, increasing water productivity deserves much attention as well. Increasing water productivity means either to produce the same yield with less water resources or to obtain higher crop yields with the same water resources [31] which would enable growing more food and hence feeding the ever increasing population or gaining more benefits with less water [32] . In its broadest sense, it reflects the objectives of producing more food, income, livelihood and ecological benefits at less social and environmental cost per unit of water consumed [33] .

Sub-Saharan Africa countries, often characterized with low yield levels and large on-farm water losses resulting in poor water use efficiencies, have a promising potential for improvements in water productivity through introduction of agronomic or engineering measures [15] [33] . An important means of achieving increased water productivity will be to get more from the amount of water that is beneficially depleted by agriculture which can be achieved by introducing different agronomic, engineering and management technologies such as changing crop varieties, crop substitution, deficit, supplemental or precision irrigation [7] [16] . The next sections discuss small holder drip irrigation and water pricing as examples of the engineering and management techniques to improve water productivity and, of course, water use efficiency.

2.1. SMALL HOLDER DRIP IRRIGATION SYSTEM

Drip irrigation's combination of water savings and yield increases typically produces at least a doubling of water productivity, yield per unit water, and makes it a leading technology in the global challenge of boosting crop production in the face of serious water constraints [34] cited in [35] . However, due to its high capital cost, particularly for the poor farmers, the conventional drip irrigation is generally not widely used on a global scale and accounts for less than 0.1% of irrigated land worldwide [36] . Nonetheless, smallholders drip irrigation system attempts to retain the benefits of conventional systems while removing the factors that prevent their adoption by poor smallholders: purchase cost, the need for a pressurized supply, the associated pumping costs, and the complexity of operation and maintenance. It can help solve water management problems faced by smallholder farmers by making it easier and simpler to supply the right amount of water to their crops at the right place and at the right time [37] . Drip irrigation used about 35% of the water used by the surface irrigation systems thus giving much higher water use efficiencies [38] . It was also concluded by [38] that low cost drip systems achieved water saving of more than 50% compared to surface irrigation systems. Further results from field trials demonstrate that affordable smallholders drip systems easily pay for themselves in one growing season, and stimulate shifts to more intensive agricultural practices by small farmers [39] . Implementation of this technology might help improve livelihoods of individual farmers or small communities and is mentioned promising in the context of poverty alleviation which is the number one enemy of many sub-Saharan Africa countries [36] .

Despite the higher potential for smallholder drip irrigation and other small scale irrigation technologies as supplementary irrigation for millions of people and to achieve household food security, the use of the technology has hardly begun in many sub-Saharan Africa countries. With proper marketing and proper agronomic and technical support low cost irrigation technologies can be adopted by smallholder farmers to change family lifestyles, increase people's incomes, create employment and go a long way towards food security and improved nutrition [38] . However, recent literatures argue that the success of the technology is interpreted through development agencies lenses and with the intention of continuing involvement in future projects [40] [41] . Hence, it is recommended to get the technology tested for its efficiency, productivity and social acceptance by farmers and independent researchers before its introduction and distribution to the vast agrarian population.

2.2. WATER PRICING

Water pricing has become a key issue in both the developed and developing world. It is designed taking into account various objectives which are oftentimes more and more difficult to achieve due to environmental constraints, increased water demand, climate change and economic crisis which impacts all incomes (households, farmers, industries). The three main objectives of water pricing are balancing water budget, allowing water access, and inciting users to save water [42] . Achieving these objectives directly leads to a better water resource management which could help us in coping with the scarce and still shrinking water resource. [20] reported that pricing of water and water services are an acceptable practice that may result in better conservation and sustainable management of water. This is because higher prices for water use would encourage its more rational use by agricultural users and provide funds for investing in infrastructure which are two necessary conditions to avoid risk of water shortage [43] .

Although, water pricing has long been practiced in some developed countries, it has been rarely or hardly implemented in sub Saharan Africa. The question here would be "how to introduce water pricing for the first time in an area where water has been accessed for free?" and "on what basis to effect the pricing?". In principle, there are multiple pricing patterns for irrigation water. The most common pricing systems are based on volume, area, electricity consumption (for groundwater based irrigation), type of irrigation or crop and block rate pricing structure [44] . The experience of the countries which have already implemented different pricing systems can be taken as a bench mark to choose better pricing approach in a particular area. The area pricing system, modified according to the crop or irrigation techniques, does not encourage water saving for a given choice of crop or irrigation technique, but it does have more effect on the choice of which crops to irrigate or which irrigation technique to adopt [45] . Pricing systems based on volume, volumetric pricing, and block rate pricing structure are known for their effect on high water savings. [46] cited in [47] estimated a 40% reduction in irrigation water demand associated with the increasing block rate structure.

Water pricing alone will not always be a sufficient incentive for users to enhance water saving and improve irrigation performance [45] . Thus, due consideration is needed to be given to water distribution and control. Rational water use has to be achieved by careful control of distribution and by allocations that broadly meet crop requirements. In this regard, it is required to enhance understanding of water scarcity among irrigators and encourage more efficient water use. It is also worthy to ensure better control over the supply of water as it can reduce wastage of water associated with excess amounts of water flowing through uncontrolled canals and un- gated turnouts onto fields and into drainage channels [48] .

3. CONCLUSION

Water scarcity is becoming severe and the availability of water for agriculture is dwindling due to increasing population and food demand, changing climate, expansion of industries and manufacturing. In order to sustain their economy and feed their population, sub-Saharan Africa countries should implement sound, feasible and implementable water management policy that can help in obtaining more production from the existing resources, land and water, through maximization of water use efficiency and crop water productivity. It is, thus, wise to develop a set of engineering and management interventions such as water-saving techniques, improved irrigation management, supplemental or precision irrigation, water pricing and better agronomy. Before implementation, however, the intervention measure should be tested through independent and participatory research. Above all, farmers, government and non-governmental organizations should cautiously evaluate the feasibility of the intervention measures and proper awareness should be given to the farming community.

REFERENCES

1. United Nations (2015) World Population Prospects: The 2015 Revision. DVD Edition, Department of Economic and Social Affairs, Population Division. United Nations, New York.
2. Rockström, J., Barron, J. and Fox, P. (2003) Water Productivity in Rain-Fed Agriculture: Challenges and Opportunities for Smallholder Farmers in Drought-Prone Tropical Agroecosystems. In: Kijne, J.W., Barker, R. and Molden, D, Eds., Water Productivity in Agriculture: Limits and Opportunities for Improvement, CABI, Wallingford, 145-162.
3. Playan, E. and Mateos, L. (2006) Modernization and Optimization of Irrigation Systems to Increase Water Productivity. Agricultural Water Management, 80, 100-116.
4. World Bank (2016) Health Nutrition and Population Statistics. World Bank, Washington DC.
5. Shiklomanov, I.A. (2000) Appraisal and Assessment of World Water Resources. Water International, 25, 11-32.
6. Molden, D. (1997) Accounting for Water Use and Productivity. SWMI Paper, International Water Management Institute, Colombo.
7. Howell, T.A. (2001) Enhancing Water Use Efficiency in Irrigated Agriculture. Agronomy Journal, 93, 281-289
8. Gleick, P.H. (2002) Water Management: Soft Water Paths. Nature, 418, 373.

9. Heumesser, C., et al. (2013) Current State and Future Potential of Global Food Production and Consumption. In: Rengel, Z., Ed., Improving Water and Nutrient-Use Efficiency in Food Production Systems, John Wiley & Sons, Inc., Hoboken,, 1-19.

10. Bruinsma, J. (2003) World Agriculture: Towards 2015/2030. An FAO Perspective. Earthscan Publications Ltd., London.

11. Bruinsma, J. (2009) The Resource Outlook to 2050. Paper Presented at the Expert Meeting on How to Feed the World in 2050, Rome, 24-26 June 2009.

12. Falkenmark, M., Lundqvist, J. and Widstrand, C. (1989) Macro-scale Water Scarcity Requires Micro-Scale Approaches. Natural Resources Forum, 13, 258-267.

13. Falkenmark, M. (1989) The Massive Water Scarcity Now Threatening Africa: Why Isn't It Being Addressed? Ambio, 18, 112-118.

14. Ghanem, H. (2008) The State of Food Insecurity in the World, 2008: High Food Prices and Food Security: Threats and Opportunities. Food and Agriculture Organization of the United Nations (FAO), Rome.

15. Kijne, J.W., Barker, R. and Molden, D. (2003) Improving Water Productivity in Agriculture: Editors' Overview. In: Kijne, J.W., Barker, R. and Molden. D., Eds., Water Productivity in Agriculture: Limits and Opportunities for Improvement, CABI, Wallingford, xi-xix.

16. Molden, D. and Sakthivadivel, R. (1999) Water Accounting to Assess Use and Productivity of Water. International Journal of Water Resources Development, 15, 55-71.

17. De Fraiture, C., et al. (2007) Looking Ahead to 2050: Scenarios of Alternative Investment Approaches. In: Molden, D., Ed., Water for Food, Water for Life: A Comprehensive Assessment of Water Management in Agriculture, Earthscan/IWMI, London/Colombo, 91-145.

18. Garnett, T.P. and Rebetzke, G.J. (2013) Improving Crop Nitrogen Use in Dryland Farming. Improving Water and Nutrient-Use Efficiency in Food Production Systems, Chap. 8, 123-144.

19. Kumar, M.D., Sivamohan, M. and Bassi, N. (2012) Water Management, Food Security and Sustainable Agriculture in Developing Economies. Routledge, Abingdon-on-Thames.

20. Dinar, A., Pochat, V. and Albiac-Murillo, J. (2015) Water Pricing Experiences and Innovations. Global Issues in Water Policy. Springer International Publishing, Gewerbestrasse.

21. Seckler, D.W. (1998) World Water Demand and Supply, 1990 to 2025: Scenarios and Issues. Vol. 19, Iwmi, Colombo, Sri Lanka.

22. Awulachew, S.B., et al. (2005) Experiences and Opportunities for Promoting Small-Scale/Micro Irrigation and Rainwater Harvesting for Food Security in Ethiopia. International Water Management Institute, Addis Ababa.

23. Rosegrant, M.W., Cai, X. and Cline, S.A. (2002) World Water and Food to 2025: Dealing with Scarcity. International Food Policy Research Institute.

24. Cooper, P., Dimes, J., Rao, K.P.C., Shapiro, B., Shiferaw, B. and Twomlow, S. (2008) Coping Better with Current Climatic Variability in the Rain-Fed Farming Systems of Sub-Saharan Africa: An Essential First Step in Adapting to Future Climate Change? Agriculture, Ecosystems & Environment, 126, 24-35.

25. Wani, S.P., Rockström, J. and Oweis, T.Y. (2009) Rainfed Agriculture: Unlocking the Potential. Vol. 7, CABI, Wallingford.

26. Rosegrant, M.W., et al. (2002) The Role of Rainfed Agriculture in the Future of Global Food Production. Environment and Production Technology Division, International Food Policy Research Institute.

27. Gleick, P.H. (2003) Global Freshwater Resources: Soft-Path Solutions for the 21st Century. Science, 302, 1524-1528.

28. Hoanh, C.T., Johnston, R. and Smakhtin, V. (2015) Climate Change and Agricultural Development: A Challenge for Water Management. In: Hoanh, C.T., Johnston, R. and Smakhtin, V., Eds., Climate Change and Agricultural Water Management in Developing Countries, CABI, Wallingford, 1.

29. Fuchs, M. (2007) Impact of Research on Water Use for Irrigation in Israel. Irrigation Science, 25, 443-445.

30. Stanhill, G. (1992) Irrigation in Israel: Past Achievements, Present Challenges and Future Possibilities. Water Use Efficiency in Agriculture: Proceedings of the Binational China-Israel Workshop, Beijing, 22-26 April 1991, 63-77.

31. Zwart, S.J. and Bastiaanssen, W.G. (2004) Review of Measured Crop Water Productivity Values for Irrigated Wheat, Rice, Cotton and Maize. Agricultural Water Management, 69, 115-133.

32. Yihun, Y.M., Haile, A.M., Schultz, B. and Erkossa, T. (2013) Crop Water Productivity of Irrigated Teff in a Water Stressed Region. Water Resources Management, 27, 3115-3125.

33. Molden, D., et al. (2009) Improving Agricultural Water Productivity: Between Optimism and Caution. Agricultural Water Management, 97, 528-535.

34. Postel, S. (1999) Pillar of Sand: Can the Irrigation Miracle Last? W.W. Norton & Company, New York.

35. Postel, S., Polak, P., Gonzales, F. and Keller, J. (2001) Drip Irrigation for Small Farmers. Water International, 26, 3-13.

36. Mehari Haile, A., Depeweg, H. and Stillhardt, B. (2003) Smallholder Drip Irrigation Technology: Potentials and Constraints in the Highlands of Eritrea. Mountain Research and Development, 23, 27-31.

37. Stillhardt, B., Ghebru, B. and Mehari, A. (2003) Small-Scale Micro Irrigation in Eritrea. A Feasibility Study on the Introduction of Affordable Micro Irrigation Technology in Eritrea. Environment (CDE), University of Bern, Hochschulstrasse.

38. Maisiri, N., Senzanje, A., Rockstrom, J. and Twomlow, S.J. (2005) On Farm Evaluation of the Effect of Low Cost Drip Irrigation on Water and Crop Productivity Compared to Conventional Surface Irrigation System. Physics and Chemistry of the Earth, Parts A/B/C, 30, 783-791.

39. Postel, S. (2000) Entering an Era of Water Scarcity: The Challenges Ahead. Ecological Applications, 10, 941-948.

40. Wanvoeke, J., Venot, J.-P., De Fraiture, C. and Zwarteveen, M. (2015) Smallholder Drip Irrigation in Burkina Faso: The Role of Development Brokers. The Journal of Development Studies, 52, 1019-1033.

41. Venot, J.-P. (2016) A Success of Some Sort: Social Enterprises and Drip Irrigation in the Developing World. World Development, 79, 69-81.

42. Montginoul, M., Loubier, S., Barraqué, B. and Agenais, A.-L. (2015) Water Pricing in France: Toward More Incentives to Conserve Water. In: Dinar, A., Pochat, V. and Albiac-Murillo, J., Eds., Water Pricing Experiences and Innovations, Springer International Publishing, Gewerbestrasse, 139-160.

43. De Andrade Resende Filho, M., Correa, J.S.O. and de Oliveira Torres, M. (2015) Water Pricing in Brazil: Successes, Failures, and New Approaches. In: Dinar, A., Pochat, V. and Albiac-Murillo, J., Eds., Water Pricing Experiences and Innovations, Springer International Publishing, Gewerbestrasse, 41-61.

44. Che, Y. and Shang, Z. (2015) Water Pricing in China: Impact of Socioeconomic Development. In: Dinar, A., Pochat, V. and Albiac-Murillo, J., Eds., Water Pricing Experiences and Innovations, Springer International Publishing, Gewerbestrasse, 97-115.

45. Chohin-Kuper, A., Rieu, T. and Montginoul, M. (2003) Water Policy Reforms: Pricing Water, Cost Recovery, Water Demand and Impact on Agriculture. Lessons from the Mediterranean Experience. Proceedings of the Water Pricing Seminar, Agencia Catalana del Agua and World Bank Institute, Barcelona, 30 June -2 July 2003, 1-9.

46. Pike, T., et al. (2007) Does Pricing Water Reduce Agricultural Demand? An Example from British Columbia. Sustainable Development Briefing Note. Government of Canada, Policy Research Initiative.

47. Renzctti, S. and Dupont, D.P. (2015) Water Pricing in Canada: Recent Developments. In: Dinar, A., Pochat, V. and Albiac-Murillo, J., Eds., Water Pricing Experiences and Innovations, Springer International Publishing, Gewerbestrasse, 63-81.

48. Molle, F. and Berkoff, J. (2007) Water Pricing in Irrigation: The Lifetime of an Idea. In: Molle, F. and Berkoff, J., Eds., Irrigation Water Pricing: The Gap between Theory and Practice, CABI, Wallingford, 1-20.

6

CHAPTER

Measuring Soil Water Potential for Water Management in Agriculture: A Review

Marco Bittelli

Department of Agro Environmental Science and Technology, University of Bologna, Viale Fanin, 44, Bologna, Italy

ABSTRACT

Soil water potential is a soil property affecting a large variety of bio-physical processes, such as seed germination, plant growth and plant nutrition. Gradients in soil water potential are the driving forces of water movement, affecting water infiltration, redistribution, percolation, evaporation and plants' transpiration. The total soil water potential is given by the sum of gravity, matric, osmotic and hydrostatic potential. The quantification of the soil water potential is necessary for a variety of applications both in agricultural and horticultural systems such as optimization of irrigation volumes and fertilization. In recent decades, a large number of experimental methods have been developed to measure the soil water potential, and a large body of knowledge is now available on theory and applications. In this review, the main techniques used to measure the soil water potential are discussed. Subsequently, some examples are provided where the measurement of soil water potential is utilized for a sustainable use of water resources in agriculture.

Keywords: soil water potential; water use; irrigation; agriculture

1. INTRODUCTION

Water resources management is one of the most pressing environmental issues, especially because of the competition between industry, agriculture, municipal and energy utilization of water resources. At the global level, 80%–90% of all the water is consumed in agriculture. Unfortunately, water use efficiency in agriculture is poor with more than 50% water losses, making possible to save enormous water quantities in the agricultural sector, when compared with other sectoral water uses [1]. Moreover, in many regions of the world, plant growth and crop yields are limited by water deficits, as more frequent and intense droughts and the areas of land characterized as very dry, have increased in recent decades [2]. Agriculture utilizes a large part of the total fresh water not only in dry but also in sub-humid environments. For instance, in the Po river basin (the largest river in Italy, supplying fresh water to millions of people), of the total water drawn from the river, 16% is used for municipal usage (drinking water), 20% is used for industrial activities (processes and cooling), 18% for energy (hydroelectric power) and 46% for agriculture (irrigation).

The most important strategies for improving water use efficiency in agriculture are: (a) improvement of the water distribution system with reduction of seepage losses in canals and pipelines, (b) optimization of irrigation techniques, (c) correct choice of cropping systems adapted to the climatic and pedological conditions of the site and (d) use of treated waste water and reutilization of irrigation waters.

For the optimization of irrigation techniques, improvement of irrigation planning and employment of specific cropping systems, it is often necessary to compute the water budget of a given farm, catchment, river basin or region. In many cases, when the soil water budget is computed, it shows that the most important term in irrigated agriculture is evapotranspiration, which can make up 50 or 60% of the total water budget of a farm [3], stressing the importance of correct irrigation management and optimization of plant's water use.

The soil water budget is affected primarily by soil water dynamics (infiltration, redistribution, percolation, evaporation, plant's transpiration), which are, among other variables, regulated by the soil energy state of water, *i.e.*, the soil water potential (SWP). Knowledge of the SWP may also be used to reduce percolation of solutes from the vadose zone into the groundwater, as well as for planning alternative cropping systems based on the plants' response to water stress [4]. Clearly, knowledge of the SWP alone is not sufficient to provide all the necessary information needed to compute the soil water budget, however it is an important variable that can be utilized to improve water use in irrigated agriculture.

SWP has been employed in a variety of studies and applications to optimize irrigation schedule and to investigate ways in which water-saving irrigation can help to save water, increase water productivity and optimize production. Bouman *et al.* (2001) [5] utilized measurements of SWP to assess the intensity of drought in irrigated low land rice. In a field study on drip irrigation on tomato, Wang *et al.* (2007) [6] studied the effect of SWP and irrigation scheduling on tomato production. Thompson *et al.* (2007) [7] pointed out that, for greenhouse-grown vegetable crops, soil matric potential thresholds are much better indicators for irrigation scheduling than available water content thresholds. Viswanatha *et al.* (2002) [8] utilized, among other measurements, the water potential to study the relationship between irrigation water and sweet corn yield.

A successful, large scale example of application of water budget models and SWP measurements to save irrigation water in commercial farms is provided by the Irrinet (IT Irrigation Advisory Services for Farm Water Management) service [9]. Irrinet is the irrigation planning system managed by the Irrigation Service of the Emilia Romagna Water Channel Agency (CER, Italy). The support system is based on a water balance model, and irrigation schedules and volumes are tuned for different crops and environments according to specific research and studies carried out locally for more than 20 years. The Irrinet service has been developed to optimize the use of water at the farm scale by transferring personalized technical indications to each farmer.

As today, the system is used on 23% of the total farmed areas of the Emilia-Romagna region. It is utilized by more than 13,000 farmers and it is allowing an yearly estimated reduction of 20% of the water utilized in irrigated agriculture. The service offers technical information provided by an internet-based system where the farmer can login and obtain personalized irrigation information. Direct measurements of soil properties such as the SWP, as well as other information (crop water requirements, crop growth, root system, crop evapotranspiration) are monitored in various farms, for calibration and optimization of the model used for the irrigation support system.

Overall, water management in agriculture plays a key role in determining effective water savings plans and optimization of natural resources for sustainable agricultural activities. It can take advantage of a variety of experimental and modelling techniques to improve its efficiency. In this review, the role and the importance of SWP for the soil water budget is first reviewed and the most important experimental techniques used to measure SWP are described. Finally, some examples of application of these techniques for water management in agriculture and their application for optimization of water resources, are presented.

2. THE SOIL WATER POTENTIAL

The SWP potential is defined as the potential energy of water in soil. The SWP is quantified relative to a standard state where water has no solutes, free from external forces except gravity, at a reference pressure, at a reference temperature and at a reference elevation [10,11]. The SWP is then defined as the energy state of water in soil with respect to the energy of water at the reference state.

The total SWP (ψt) is determined by a variety of forces acting on the soil water, including gravitational (ψg), matric (capillary and adsorptive, ψm), osmotic (ψo) and hydrostatic (ψh).

$$\psi_t = \psi_g + \psi_m + \psi_o + \psi_h \tag{1}$$

The gravitational potential is determined by the elevation of the soil water in respect to the chosen reference elevation, which is usually the soil surface. The matric potential is determined by the forces exerted by the soil matrix (soil particles) on water, which are capillary and adsorptive forces. The osmotic potential is given by the presence of solutes in the soil solution, which decrease the potential energy of water. The hydrostatic potential (also called pressure potential) is determined by a pressure exerted by overlaying water over a point of interest in the soil. Other components of the water potential can also be defined.

Since the SWP represents an energy, it should be expressed as a unit of energy per unit mass (J kg-*1*) or per unit volume of water (J m-*3*). The former is preferable because it does not require to include in the computation the changes of water volume with temperature. However, SWP is often expressed in related units such as energy per unit weight, which is the equivalent of a head of water. The energy is equivalent to the pressure exerted by a liquid column of a given height. For instance, a column of water of ~10 meters correspond to a pressure of 1 bar. This unit is more common because it is simpler and it allows for easy visualize of the gravitational potential and the pressure potential, which are often expressed in meters.

Another common unit is the pressure (Pa or bar), derived from the first methods developed to measure the soil water potential such as the tensiometers and the pressure plate apparatus [12,13]. This unit is based on the definition of energy per unit volume. Since pressure in a fluid can be seen to be a measure of energy per unit volume, energy is given by the product of pressure by volume (E = P × V). The water potential can then be expressed as a pressure and there is an equivalence with the definition of energy per unit volume. It is useful to remember that there is a direct numerical equivalence between the units of J kg-*1* and kPa.

Usually the SWP is expressed as a negative number because it represents the energy required to transfer the soil water to the reference state described above. The term "suction" and "tension" are definitions developed to avoid using the negative sign and to represent the SWP as positive numbers. However, in this review, the negative sign will be retained and the property will be called water potential.

The SWP can range over several orders of magnitude, from a few Joules per kilogram when the soil is close to saturation to minus thousands of Joules per kilogram when the soil is very dry. For agricultural applications usually the range of interest is between 0 and approximately −1,500 J kg-*1*, the latter being defined as the soil water potential at the permanent wilting point of many cultivate crops. However, measurements at more negative water potentials may be needed for water studies in dry and very dry climate.

As at today there is not a single technique able to measure the SWP over the entire range of interest and this limitation is one of the most important challenges.

3. THE SOIL-PLANT-ATMOSPHERE CONTINUUM

The movement of water from the soil, into the roots, through the xylem and from the leaf into the atmosphere, occurs because of a series of water potential gradients. The system that involves the soil, the plant's roots, the xylem, the leaf and the atmosphere, is called the soil-plant-atmosphere continuum (SPAC), which is a pathway for the movement of water from the soil into the atmosphere. Figure 1shows a schematic of the soil-plant-atmosphere continuum. On the right side of the figure is indicated the water potential of the different elements of the continuum (Ψ) and the resistance terms (R), while on the left side a schematic of the different water movements is depicted.

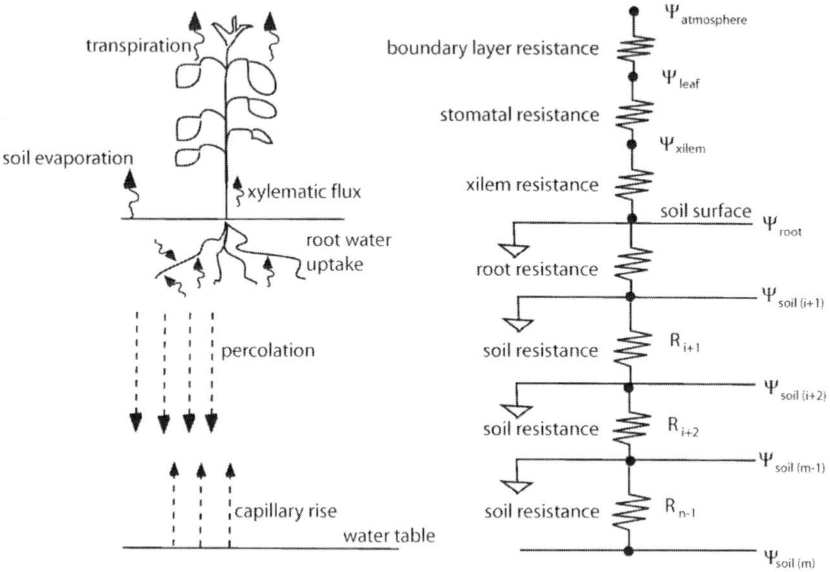

Figure 1. Schematic of the soil-plant-atmosphere continuum.

Water flows from soil to roots, through xylem, mesophyll and wall cells, and it evaporates through the substomatal cavities. From the substomatal cavities it diffuses out of the stomata, through the leaf and the canopy into the atmosphere. The value of the water potential is higher (less negative) in the soil and decreases along the transpiration pathway. The water potential values in the different elements of the system (saturated and unsaturated soil, plant's roots, plant's xylem, plant's leaves, atmosphere) are determining a series of water potential gradients that are the driving forces for water movement.

Under a given gradient, the magnitude of the flux depends on the value of the resistance terms along the pathway. According to Campbell (1985) [14], within the plant the xylem resistance is negligible compared to other resistances. The major resistance are the endodermis, where water enters the root steele and in the leaf in the bundle sheath. Usually, in the plant the potential drops occur for 60% to 70% in the endodermis and 30% to 40% in the leaf.

To understand the driving forces that determine the movement of water through the SPAC, it is useful to describe the relationship between water potential and relative humidity at a liquid-air interface, the Kelvin equation:

$$\psi = \frac{RT}{M_w} \ln h_r \qquad (2)$$

where ψ is the water potential (J kg-*1*), R is the gas constant (8.3143 J mol-*1* K-*1*), T is temperature (K), Mw is the molecular weight of water (0.018 kg mol-*1*), and hr is the relative humidity (expressed as a fraction from 0 to 1). Equation 2 can be inverted:

$$h_r = \exp \frac{M_w \psi}{RT} \qquad (3)$$

to obtain the relative humidity from water potential.

For instance, according to Equation 2, the atmosphere at 0.6 of relative humidity, at 20 ◦C, has a water potential of 69,133 (J kg-*1*). For a loamy soil, at 20◦C and at field capacity (which is the soil water content after water in the largest pores has drained away and the rate of downward movement has become small), the water potential is about −33 (J kg-*1*), and according to Equation 3 the corresponding relative humidity (*hr*) at the liquid-air interface is 0.9997. This large difference in water potential between the soil and the atmosphere is the driving force for water movement from the soil into the atmosphere.

When water is uptake from the soil into the roots, it moves into the xylem and into the leaves and it undergoes a series of changes that reduce the water potential, but the changes are not as dramatic as they are between the soil (or the plant) and the atmosphere. The SWP usually ranges from values close to zero (when the soil is at or close to saturation) to values of −1,000 to −2,000 (J kg-*1*) when the soil is very dry. The leaf water potential usually ranges from −100 (J kg-*1*) for a well watered plant to −2,000 (J kg-*1*) for a water stressed plant [15].

These values of water potential shows that, in the SPAC, the large water potential gradient occurs between the leaf and the atmosphere (when the atmosphere relative humidity decreases), and it is precisely this gradient that determines plant transpiration and it supports the water cycle as we know it. Moreover, the change of phase between liquid and vapor occurring between the leaf and the leaf boundary layer requires a relatively high amount of energy (the latent heat of vaporization), which determine a plant's decrease of temperature and it is a key phenomena for the plant's thermal regulation. The transpiration process (regulated by the plant through the stomata) has therefore also a key role in the plant's thermal regulation.

Knowledge of soil and plant water potential is hence of utmost importance for soil water availability itself and to quantify plants' water requirements. Indeed, the concept of soil water content (or soil wetness) is not sufficient and sometime misleading, to describe the availability of soil water for plants, because it is otherwise the energy state of water that determines the flow of water in the soil-plant-atmosphere continuum. However, the relationship between soil water content and water potential is an important property that is described in the following section.

4. THE SOIL WATER RETENTION CURVE

When a soil is completely saturated (*i.e.*, the pore space is entirely filled with water), at equilibrium with free water, and at the same height, the potential is zero, because the gravitational, the matric and the hydrostatic pressure are zero. If the pressure is changed (for instance by increasing the elevation in respect to the free water, or applying an external pressure) a negative potential is determined in the soil sample and the pores start to desaturate (water leaves the soil sample).

The largest pores are the one that desaturates first and, at increasingly lower water potential values, progressively smaller pores desaturates. When the first largest pore desaturates, the value of the corresponding water potential is called air entry potential, since it is the value corresponding to the initial entering of air into the saturated soil.

As the pores desaturate, the total soil water content of the soil sample decreases. The relationship between the water potential and the corresponding values of soil water content is called the soil water retention curve (WRC). Figure 2 shows an example of three WRC for soils having different textural composition. The Salkum sample (silty clay loam) is the soil with the largest water holding capacity, followed by the Walla Walla (silt loam) and the L-Soil (sand) samples, showing that the shape of the WRC depends primarily on texture. Usually, the largest the clay content the greater the water holding capacity, while the largest the sand content the smaller the water holding capacity, because sandy soils are mostly characterized by large pores that quickly desaturate, leaving a small amount of water at lower water potentials. The depth of the soil profile has also an important role in determining the soil water holding capacity.

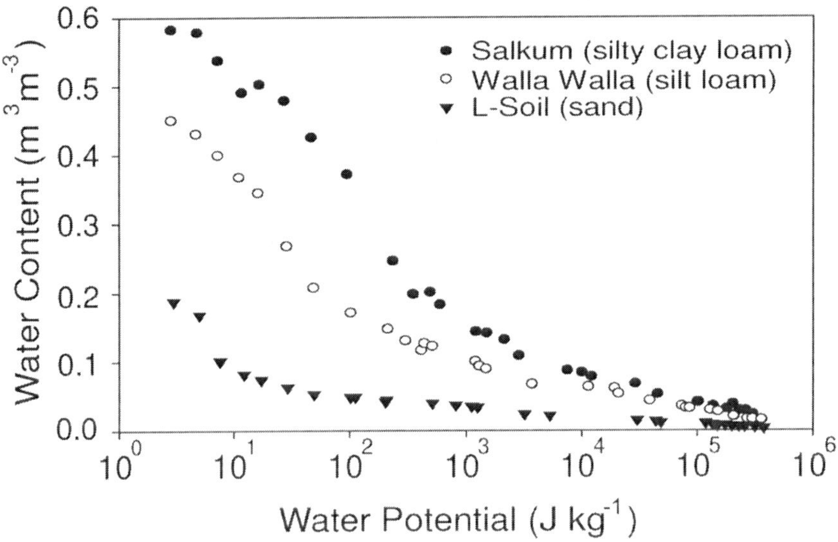

Figure 2. Soil water retention curves for three soils having different texture (From Campbell and Shiozawa, 1992).

The soil WRC, together with the the hydraulic conductivity curve (relating the soil water potential and the unsaturated hydraulic conductivity) are the most important soil hydraulic properties. They affect the soil water budget, since it affects the ability of the plant's roots to uptake water from the soil, the amount of water percolating into the groundwater, the microbial decomposition and therefore the amount of soil organic matter, as well as the solute concentration and therefore phenomena such as soil fertility, salinization and aridity. In the following sections, different experimental techniques used to measure the SWP (as well as the WRC) are described.

5. SOIL WATER POTENTIAL MEASUREMENT: GENERAL PRINCIPLES

The general principle for SWP measurement is *hydraulic equilibrium* between the water held in the soil and the measuring device. Equilibrium can be reached for the liquid phase or for the vapor phase. Tensiometers (one of the most common devices) are based on the equilibrium between liquid water in soil and liquid water in a porous cup. The small pores of the porous cup allows the establishment of an*hydraulic continuum* between the tensiometer water reservoir and the soil. The pressure changes determined by the differences between pressure inside the tensiometers and the soil are then detected by a pressure sensor.

Other instruments, such as heat dissipation sensors or dielectric sensors are also based on the equilibration of a ceramic cup with the surrounding soil. In this case, however, a property of the ceramic (*i.e.*, thermal conductivity, dielectric permittivity) that depends on the ceramic water content is measured. Since there is a relationship between the ceramic water content and the ceramic water potential (the ceramic's water retention curve) it is possible, through calibration curves, to obtain the soil water potential of the ceramic's pore water. Under the assumption of *hydraulic equilibrium*, the water potential of the ceramic cup should be equal to the one of the surrounding soil.

Other instruments, such as thermocouple psychrometry or dew point potential meters are based on the equilibrium of the vapor phase. According to Equation 2, the water potential in the air phase is related to relative humidity. Therefore, the measurement of relative humidity and temperature of air in equilibrium with soil allows for obtaining the water potential.

Thermocouple psychrometry is based on temperature difference between a dry bulb and wet-bulb temperature. When the relative humidity of the vapor phase is low, the vapor pressure deficit will be high, the evaporation rate will be high and therefore the temperature depression will be higher. In the dew point methods, the saturation vapor pressure is first measured by a thermometer, the temperature of a close-chamber is decreased until dew forms on a chilled mirror. The dew-point temperature can be then used to obtain the relative humidity.

Different classification schemes could be followed to classify these techniques, such as the physical principles behind the technique or the area of application. In this review the classification of SWP measurement is based on *field methods* and *laboratory methods*. This classification is useful for practical applications, however some sensors can be used both in the laboratory and in the field. In this case, the method is described only once in the section *field methods* and the "double use" is specified in each sensor's description.

6. SOIL WATER POTENTIAL MEASUREMENT: FIELD METHODS

Soil water potential in the field can be measured by employing different methods. The choice of the appropriate methods may depend on a variety of factors including:

1. Application (irrigation management, field monitoring, research)
2. Soil type (texture, bulk density, salinity, shrinking-swelling properties)
3. Meteorological conditions (high and low temperatures, freezing, humidity)
4. Location (remoteness, accessibility for sensor maintenance, remote transmission of data)
5. Cost
6. Range, accuracy, repeatability, spatial resolution
7. Technical knowledge required to operate the sensors (operator skills)
8. Required sample frequency (automatic sampling, datalogging)

The "weight" of each of these factors on the final choice must be evaluated case by case, however a short list of advantages and disadvantages have been provided for each sensor, to help the choice of the most appropriate one. Moreover, important factors such as range, accuracy, repeatability and spatial resolution have been described and discussed.

6.1. Piezometers

Piezometers are wells of small diameter, utilized to measure the hydraulic head (or elevation of the water table) in aquifers. They are usually pipes of plastic or metallic material, inserted by drilling boreholes into the ground. The piezometric measurement is based on the Bernoulli's law, which allow to compute the piezometric pressure at a point, based on the law of conservation of energy. In saturated material, with sufficient hydrostatic pressure, water will enter the pipe and equilibrium will be reached between the pipe and the aquifer.

A detailed description of the different piezometers and their applications are provided by Young (2002) [17]. The use of piezometers is widespread and there are many applications where piezometers provide very important information. For instance, measurements of the position of the water table (and its fluctuations with time) are needed in agricultural applications, for designing drainage systems and for the control of salinization. In hydrogeology and hydrology, piezometers are used for measuring hydraulic gradients in the saturated zone and to monitor groundwater contamination from point and diffuse pollution. Piezometers are also used for water sampling aimed at analyzing temperature, electrical conductivity and concentration of a variety of solutes.

6.2. Tensiometers

Tensiometers are porous ceramic cups, connected to a pressure sensor (manometer or pressure transducer) through a water-filled tube. The water-filled tube is placed into contact to the soil through a rigid porous membrane. The membrane allows water movement through the device, but not air movement. The cup is placed into the soil and the pores in the ceramic cup reach equilibrium with the soil water matric potential of the surrounding soil. When the matric potential of the soil is lower (more negative) than in the tensiometer, a water potential gradient is established and water from the tensiometer is attracted by the surrounding porous media, creating a suction that is detected by the pressure sensor. The pressure is read by the pressure sensor, which indicates the value of the soil matric potential.

There are a variety of tensiometers classified based on their design and on the measuring device used to sense the pressure changes. A common classification is between mechanical and electronic tensiometers. In the first ones the change in pressure is detected by a mechanical pressure gauges, while in the second ones the change in pressure is detected by en electronic pressure transducers. Usually, a membrane is deformed by a pressure change and a pressure transducer is connected to the membrane. The use of electric pressure transducers allow to obtain more precise reading and to collect the data through a datalogger, therefore allowing multiplexing and automation. Recently, advanced transducers are installed directly into the porous ceramic cup [18]. Another advancement proposed by a manufacturer (UMS, Munich, Germany) is to use a self-refilling tensiometers.

The range of tensiometers is restricted by the air entry pressure of the porous membrane, which can be derived by the well-known Young-Laplace equation that relates the water potential to the pore radius. The pore size of the ceramic material may change, although common ceramics used for tensiometers have maximum pore radius in the range of a few microns (2–3 μm). According to the manufacturer [19] the most recent tensiometers can measure water potential from 0 to 200 J kg-*1*. Other tensiometers that could reach further water potential were described by Tamari*et al.* (1993) [20]. The increase in the tensiometer range can be achieved by employing a series of manufacturing techniques. For instance, the surfaces of the ceramic should be strongly hydrophilic, the air entry value should be low-enough, and not air bubbles or cavities should be present in the device. These features are dependent on the manufacturing techniques and the choice of the materials used to build the device.

A common limitation for the use of tensiometers is the formation of air bubbles (cavitation) in the water-reservoir (*i.e.*, the water-filled tube). This phenomena occurs when the water potential decreases and the liquid water pressure inside the tube approaches its vapor pressure (at the temperature of the tensiometer), determining spontaneous evaporation and formation of air bubbles. Indeed, the lower limit described above (range) is determined by the issue of formation of air bubbles. This problem usually require refilling of the tensiometer. As mentioned above, a manufacturer (UMS, Munich, Germany) recently designed a self-refilling tensiometer.

Tensiometers may have a lag in SWP reading in respect the soil condition because of the hydraulic resistance of the cup or limited contact between the cup and the soil. This problem is particularly pronounced in swelling soils where swelling and shrinking may create air gaps between the cup and the surrounding soil, increasing the resistance to water flow and creating a lack of *continuum* between the liquid phase in soil and in the ceramic cup. In general, the reading depends on the sensitivity of the device and the resistance of the ceramic cup to water flow. Modern tensiometers, equipped with small water reservoirs and precise electronic pressure transducers have partially addressed this issue and in condition of good contact between the sensor and the soil, the response time can be in the range of a few seconds. However, since the resistance to flow changes both with the resistance of the ceramic cup and the resistance of the surrounding soil, in increasingly drier conditions the response time can become much longer due to increasing resistances to water flow.

Tensiometer reading may change depending on the size of the ceramic cup, providing different SWP measurements. According to Hendrickx *et al.* (1994) [21] an increase in tensiometer cup size reduced the variability of SWP measurements. Specifically, they suggested that the typical cup size of 42.3 cm*2* will determine a relatively large SWP measurement variability, therefore larger cup sizes should be used for field monitoring.

Figure 3 shows three tensiometers (from the bottom): (a) a traditional tensiometer with a manual manometer, (b) a tensiometer equipped with a pressure transducer and a syringe for refilling and (c) a smaller tensiometer also connected to a pressure transducer. The smaller one can be used also in the laboratory, in small vases and in greenhouses.

In general, tensiometers have the following advantages: (a) direct measurement, (b) easy to use, (c) generally inexpensive, (d) continuous reading using pressure transducers and dataloggers, (e) low skills requirement and (f) not affected by salinity. The disadvantages are: (a) limited range of SWP, (b) may take long equilibration time, (c) cavitation (air bubbles form in the water tube) preventing the reading at low water potentials.

6.3. Heat dissipation sensors

Heat dissipation sensors are porous ceramic cups, with an embedded heating element (usually a resistance) and a temperature sensor (usually a thermocouple). The ceramic cup is placed into the soil and the pores of the ceramic cup reach equilibrium with the surrounding soil. The heating element is heated by running a current through the element for a specified time interval (*i.e.*, 60 seconds) and the changes in temperature (due to heating and cooling) are measured by the temperature sensor (Figure 4). The time dependence of the temperature changes in the ceramic cup is due to its thermal conductivity, which in turn depends on its water content. Since the water content is related to the water potential of the cup (through the ceramic's water retention curve) it is possible, through calibration curves, to obtain SWP from temperature measurements [22]. Since the thermal conductivity does not display significant changes with the solute concentration of the soil solution, the measurement is not affected by soil salinity.

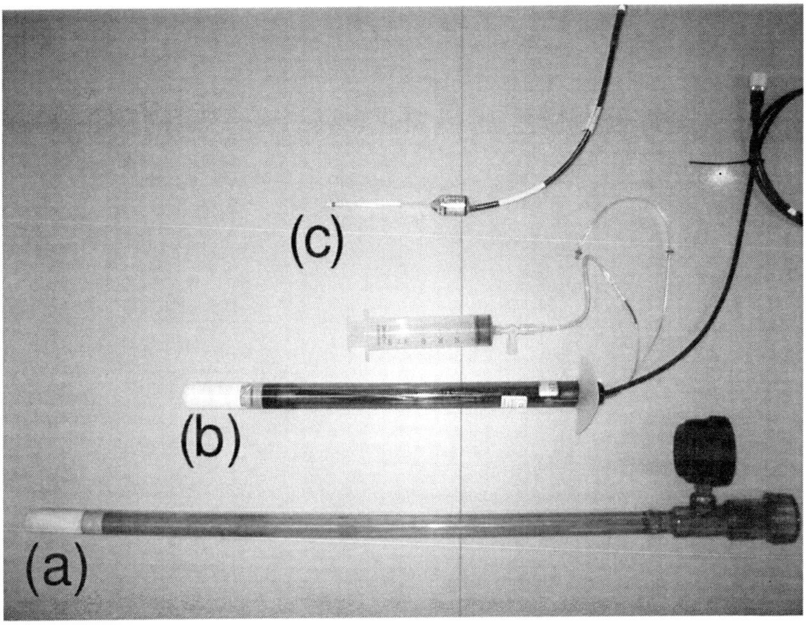

Figure 3. Tensiometers (from the bottom): (a) traditional tensiometer with manual manometer, (b) electronic large tensiometer (Model T8, UMS Inc.) and (c) electronic small tensiometer (Model T5, UMS Inc.) equipped with pressure transducers. The syringe is used to refill the T8 tensiometer.

Heat dissipation sensors measure soil water potential in the range of approximately −10*1* to −10*6* J kg-*1*, providing a wide range of water potential measurements. The upper limit is given by the air entry value of the ceramic cup, which is approximately −10*1* J kg-*1*. The accuracy is about ±1 J kg-*1* at matric potentials smaller than −100 J kg-*1*, however the repeatability is usually a limitation. Since the ceramic cups have often different pore size distributions, sensors need to be individually calibrated before installation, otherwise the sensor output would differ from one another and the repeatability of the measurement would be limited. However, according to Flint *et al.* (2002)[22], a calibration in air and pure water is sufficient without the need for a soil specific calibration, simplifying the calibration procedure. This type of sensor can be used both in the field and in the laboratory.

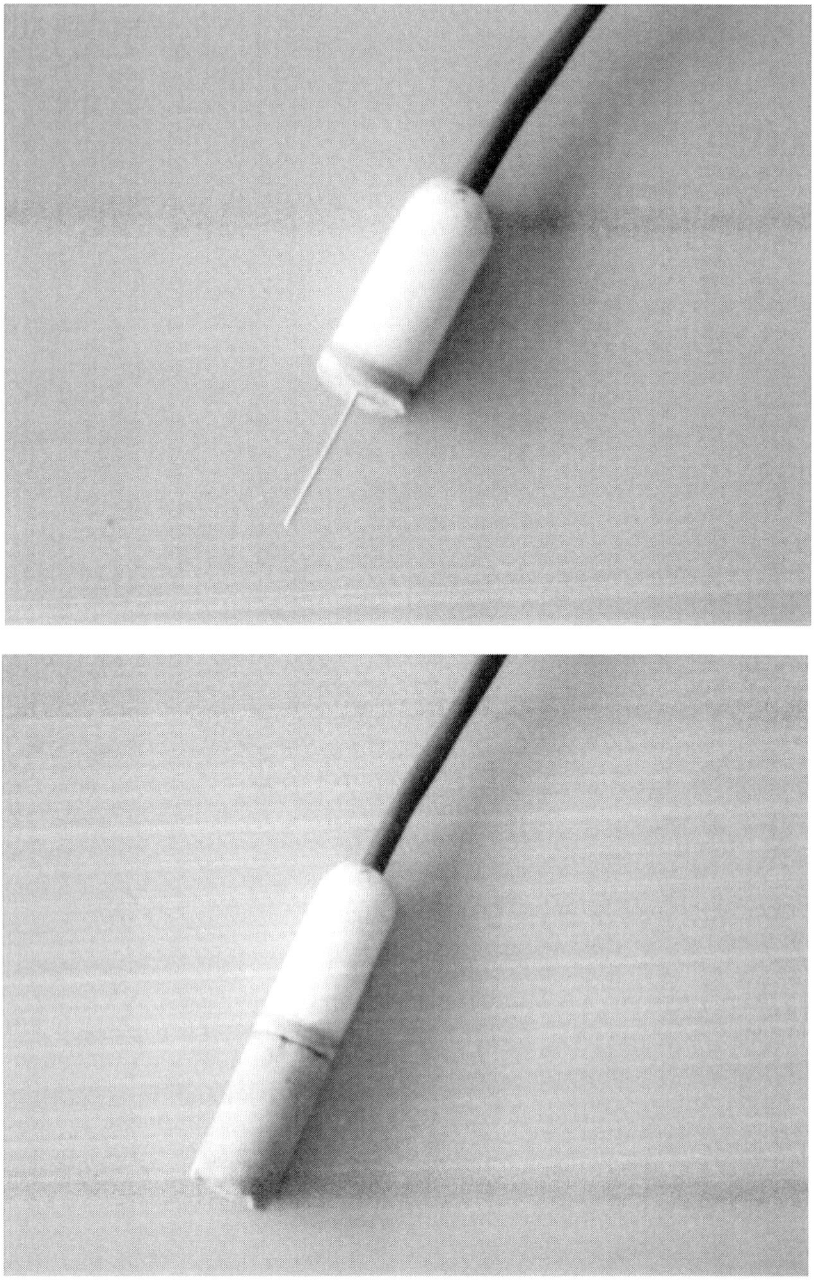

Figure 4. Heat dissipation sensor, showing the needle (the heating element and thermocouple are inserted in the needle).

In general, heat dissipation sensors have the following advantages: (a) large range of SWP, (b) lack of cavitation, (c) continuous reading using dataloggers, (e) not affected by salinity, (f) sensors are inexpensive. The disadvantages are: (a) limited upper range of SWP close to saturation, (b) sensors require individual calibration, (c) require technical skills.

6.4. Dielectric sensors

Dielectric sensors are porous ceramic cups that are placed into the soil and the pores in the ceramic cups reach equilibrium with the surrounding soil (similarly to the heat dissipation sensor). However, for these sensors the dielectric properties (dielectric permittivity) of the ceramic cup are measured, instead of the thermal properties. The dielectric properties can be measured either in the time or frequency domain, or by measuring the dielectric capacity of the porous cup. The dielectric properties are employed to measure the ceramic water content, while the water potential is inferred from calibration curves based on the relationship between water content and water potential of the ceramic porous media.

The first application was proposed by Or and Wraith (1999) [23], where soil water content was measured with TDR (Time Domain Reflectometry), and SWP was derived from calibrations. The instrument described by the authors was developed for research purposes and, to the author's knowledge, it is not commercially available. A commercial version of this type of sensor is the Matrix Potential Sensor (Model MPS-1, Decagon Dev. Inc. Pullman, WA). The sensor is made by a porous ceramic cup, surrounded by two metallic plates (forming a dielectric sandwich) that transmit an electromagnetic field through the cup (Figure 5). The metallic plates are connected to a circuit for measurement of the dielectric properties, which are highly dependent on its water content and the SWP potential is obtained from calibration curves.

According to the manufacturer, the MPS-1 range is −10 to −500 J kg-*1*. The accuracy is within 40% of the SWP measurement, while the resolution is ±1 J kg-*1*from 0 to −100 J kg-*1* and ±4 J kg-*1* from −100 to −500 J kg-*1* [24]. Also in this case, the upper range is determined by the air entry value of the ceramic.

The MPS-1 still displays problems of calibration when compared to other sensors such as tensiometers [25], due to lack of homogeneity among different ceramic cups. Moreover, there is significant hysteresis effects on the sensors. In general, similarly to the tensiometers and the heat dissipation sensor, the sensor requires good contact between the ceramic and the surrounding soil. Sometime, it may be is advisable to pack the

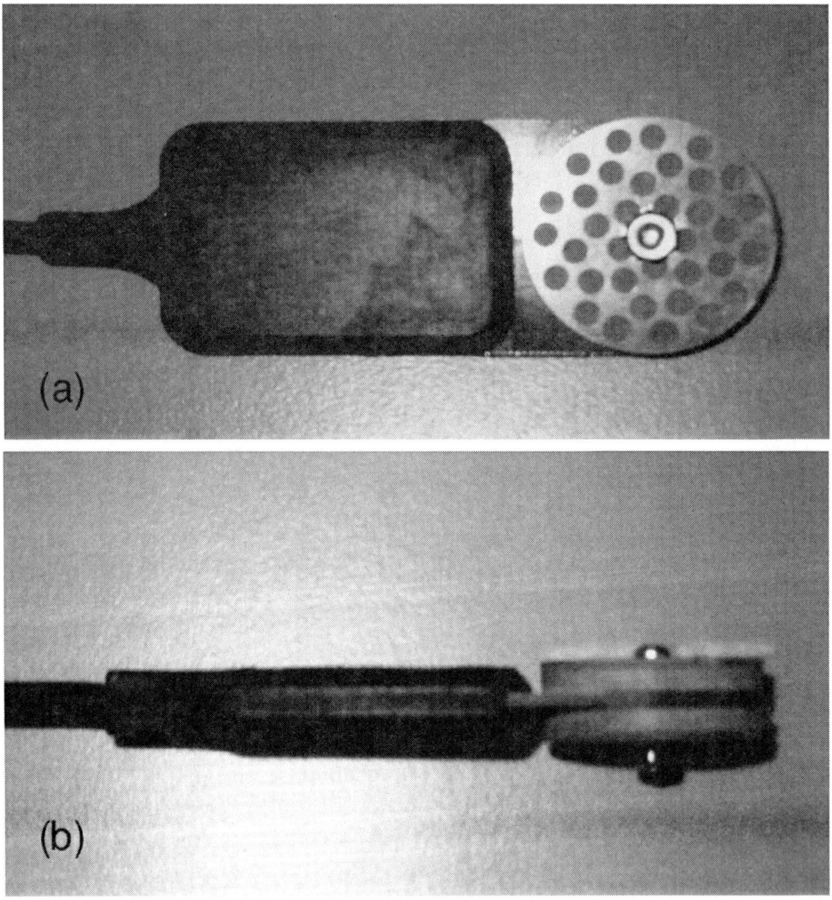

Figure 5. Matric potential sensor (MPS-1) based on the measurement of the dielectric capacity, picture from above (a) and from the side (b).

native soil around the sensor, before placing the sensor into the soil, to avoid problems of contact between the sensor and the soil.

Overall, the advantages are: (a) high accuracy, (b) provide reliable SWP readings over a very wide range, (c) allows for datalogging and remote acquisition, (d) does not require maintenance. The disadvantages are: (a) sensors needs individual calibration, (b) measurements in the wet range are limited by the air-entry value of the ceramic.

6.5. Thermocouple psychrometry

Another field (and laboratory) technique that have been widely used for measuring SWP, is thermocouple psychrometry. This technique has also been used for in situ measurement of leaf water potential [26]. Thermocouple psychrometry is a technique that derives the SWP from measurement of the relative humidity of the air phase in equilibrium with the liquid phase [27], as described in the Kelvin equation (Equation 2). The relative humidity measurement is obtained from measurements of the temperature difference between a dry bulb and wet-bulb. When the relative humidity of the vapor phase is low, the evaporation rate will be high and therefore the temperature depression will be higher. Measurements of a wet thermocouple in respect to a dry thermocouple is performed. The relative humidity is obtained from the change in temperature:

$$\frac{p}{p_0} = 1 - \left[\frac{s + \gamma^*}{p_0}\right] \Delta T \tag{4}$$

where $p/p0$ is the ratio of vapor pressure and saturation vapor pressure (the relative humidity), s is the slope of the saturation vapor pressure,γ^* is the apparent psychrometer constant and ΔT is the wet-bulb temperature depression.

The first devices for thermocouple psychrometry were built by Spanner [28] and Richards and Ogata [29]. These sensors were laboratory devices (also called sample chamber) made of a sealed chamber where the soil sample is positioned. A metal thermocouple is placed against the sample holder and the chamber is sealed, to avoid any exchange of air with the surrounding. A variety of devices equipped with a sample chamber have been developed including the Model C-52 by Wescor (Wescor Inc., Logan, Utah, USA) and the Tru-Psi Sensor by Decagon (Decagon Devices Inc., Pullman, WA, USA). The measurement range of these devices may change depending on the material used to build the sample chamber. In general, the material used to build the chamber should allow a fast thermal equilibration and, at the same time, it should not absorb water.

In general, the range and accuracy of this technique depends on the accuracy of the temperature measurement and the ability of the measuring device to keep a thermodynamic equilibrium within the sample chamber. The main limitation of this type of techniques (the dew point methods present the same limitation) is intrinsically described by the Kelvin equation, which is a logarithmic equation, where a very small change in relative humidity corresponds to a large change in water potential. For instance, at 293 K a change in relative humidity from 0.999 to 0.995, corresponds to a change in water potential from −130.5 to

−670.8 J kg-*1*, respectively. To measure changes in water potential to the accuracy of hundreds of Joules per kilogram, the measurement of relative humidity must be accurate to the third decimal. In recent years, modern microelectronics and engineering techniques allowed to improve the accuracy of this technique, through high accuracy temperature sensors and better sample-chamber designs.

The field application of *in situ* thermocouple psychrometry consists of a chamber (usually formed by a screen-cage or by a shield made by a porous ceramic cup), with the thermocouple junctions inside. Usually the wet junction is remotely controlled by a Peltier system that cools off the junction. The chamber, that is placed into the soil, allows heat and water vapor exchange with the surrounding soil [27]. As pointed out by Brown (1970) [30], if the water potential changes are slow, the sample chamber equilibrates with the soil vapor phase even if the contact between the chamber and the soil is not optimal, therefore thermocouple psychrometry should suffer to a less extent to the problems of lack of contact typical of tensiometers, heat dissipation sensors and dielectric sensors. These results were expected since these sensors are based on the equilibrium of the vapor phase and not of the liquid phase.

The type of material and the electronics used to build the chamber may affect the measurement [27,31]. For instance, McAneney *et al.* (1979) [32] reported that water potential measurement using thermocouple psychrometers built with ceramic chambers may be affected by salt adsorption in saline soils.

Overall, the advantages are: (a) high accuracy, (b) provide reliable SWP readings in very dry soil, (c) they suffer less, in respect to other methods, of problems of sensor-soil contact. The disadvantages are: (a) low accuracy at WP close to saturation (wet range), (b) slow response time, (c) small sensing volumes, (d) may be affected by soil salinity, (e) sensors require individual calibration, (f) the sensors requires accurate cleaning before calibration and use.

7. SOIL WATER POTENTIAL MEASUREMENT: LABORATORY METHODS

7.1. Soil water retention

As described above, one obtains a SWR curve through the combination of soil water potential and soil water content measurements. Therefore, the

SWP potential measurement is necessary to obtain a SWR curve. Many methods are available to determine the SWR curve, including pressure plates apparatus, thermocouple psychrometry, heat dissipation sensors, and dew point potential meters. Among these methods, the most commonly used are: (a) hanging water columns [33], (b) pressure plate apparatus [12,13,33], (c) suction tables [34] and (d) vapor pressure methods [35].

7.2. Hanging water columns

This technique is used to determine a SWR in the wet range. The sample is usually saturated and placed onto a porous cup that is in direct contact with water [36]. At the beginning of the experiment the gas pressure in the soil sample is at atmospheric pressure. A reference bottle or burette (connected to the water under the porous plate) is lowered by lowering the level *zl*, and the reduction in the water potential induces a drainage of water from the soil sample (Figure 6). The soil sample is then removed from the sample holder, weighed and oven-dried for gravimetric determination of soil water content. The procedure is repeated at further values of water potential. The method is simple, however the limitations are given by the small differences in elevation that is achievable in a laboratory (usually smaller than −2 m).

7.3. Suction plates

The suction tables [34,37], employees the same theory of the hanging columns, but the soil samples are place onto a suction plate. The suction plate is usually formed by one or more layer of fine material. The particle size distribution of the draining material determines the air entry potential, which is the maximum potential that can be imposed. The most common materials are fine sand, that determined a maximum potential of −1 to −10 J kg-*1*, and kaolinite that allows to impose smaller potential up to −50 J kg-*1*. The advantages of the suction tables in respect to the hanging columns is that they allow to measure many soil samples at a time for the same water potential. A detailed description of the method is provided by Romano *et al.* (2002) [37].

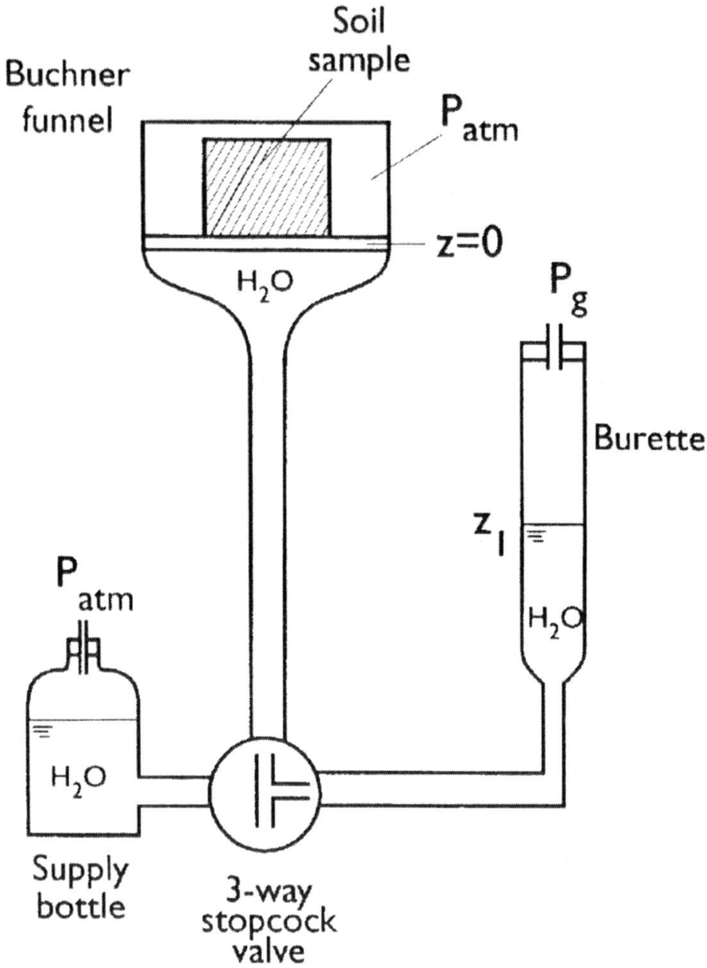

Figure 6. Schematic of the hanging water column. Scheme from Dane and Hopmans (2002).

7.4. Pressure plates apparatus

Pressure plate apparatus are closed, metal chamber where soil samples are placed onto a ceramic porous plate (Figure 7). Samples are wet on the ceramic plates to achieve good contact between the soil and plate and sometime they are covered with a polythene or cloth sheet to reduce evaporation. Usually the plates are soaked in distilled water overnight

Figure 7. Picture of the pressure plates (Soil Physics Laboratory, Department of Agro Environmental Science and Technology, University of Bologna, Italy).

before being loaded with soil samples. The pressure chambers are then closed and a specified pressure is applied by a gas tank or by an air compressor. The sample starts losing water that moves through the porous plate and is collected outside the pressure chamber. After the water ceases to drain, the samples are collected since the sample is considered in equilibrium at the specified pressure (the water potential). The soil sample is then removed from the plates, weighted and placed into an oven, for gravimetric determination of soil water content. Repeated measurements at increasing pressures are performed to obtain several measurements of water content as function of the applied pressure (water potential).

Several authors [38,39,40] showed that measurements with pressure plates are in error at potentials below −100 to −500 J kg-*1*. The authors suggested that alternative methods, such as thermocouple psychrometer or dew point other than pressure plates, are also be required to measure equilibrium water potentials at −1,500 J kg-*1*. Indeed, to avoid these problems, Campbell and Shiozawa [16] used a combination of three different techniques to determine the water retention curve: hanging water columns and pressure plates in the wet range (hanging columns from 0 to −1 m-H*2*O, and pressure plates from −10 to −500 J kg-*1*) and psychrometry in the dry range (water potentials <−500 J kg-*1*). Currently, it is advised to measure the SWR by using a combination of technique as described by Campbell and Shiozawa [16] and by Bittelli and Flury [40].

7.5. Dew point methods

The measurement of water potential can be achieved by equilibrating the soil sample in a closed chamber with the surrounding atmosphere and deriving the SWP by measuring the relative humidity of the chamber, as shown in Equation 2.

The dew point method is based on the measurement of the dewpoint temperature (Figure 8). A sample is placed into a sealed chamber. In the sealed chamber there are the following: a dewpoint sensor, an infrared thermometer and a fan. The fan is used to speed equilibrium by controlling the boundary layer conductance. The temperature of the sealed chamber is decreased until dew forms over the chilled mirror. When the dewpoint sensor detects the formation of dew, the temperature sensor measures the dew point temperature of the air, while the infrared measures the temperature of the sample. From these measurements it is possible to compute the relative humidity of the air in equilibrium with the sample.

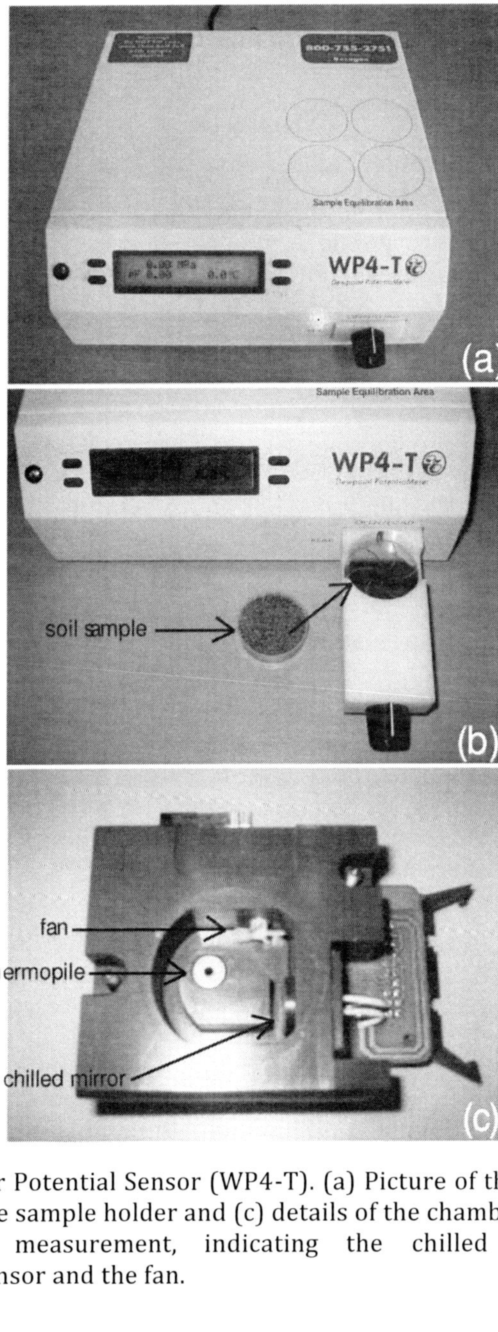

Figure 8. Water Potential Sensor (WP4-T). (a) Picture of the instrument, (b) details of the sample holder and (c) details of the chamber for relative the humidity measurement, indicating the chilled mirror, the temperature sensor and the fan.

This method has the following advantages: (a) is fast, and (b) is accurate and it covers a wide range of water potentials down to very negative values. The major disadvantages are: (a) the accuracy of the methodology decreases at water potential values close to zero and (b) the measurement is affected by soil salinity.

As described for the thermocouple psychrometers, the first limitation is simply due to the logarithmic form Equation 2 used to derive the water potential from relative humidity. The second issue is inherent of the Kelvin equation where the vapor pressure is affected by the water osmotic potential. This issue can be also an advantage since the instrument can be used to measure the osmotic potential of solutions.

Moreover, in recent years, modern microelectronics and engineering techniques allowed to improve the accuracy of chilled mirror techniques, through high accuracy temperature sensors, coupled with temperature-controlled sample chambers. Details on the measurement technique and directions to attain the best accuracy are provided by Scanlon *et al.* (2002) [35].

7.6. Freezing apparatus

A different approach to measure the WRC is based on the freezing point depression of soil water and its relationship with the SWP. It was experimentally demonstrated that the freezing point depression in porous media depends on the water potential [41]. When soil temperature drops below the freezing point of free water, the water closest to the solid particles may remain in liquid form, because the energy status of water is lowered by the forces (adsorptive and capillary) exerted by the soil, resulting in freezing point depression.

A freezing characteristic curve is obtained when soil temperature and unfrozen soil water content are measured. The water potential can be obtained from the temperature measurements by employing the Clapeyron equation and, combined to a water content measurement, allows to obtain a soil moisture characteristic (soil water retention curve). Usually, a good agreement is found between moisture characteristics obtained from freezing characteristics and moisture characteristics obtained from independent measurements with other techniques (such as pressure plates and dew point methods). For instance, in a field study Spaans and Baker (1996) [42] obtained good agreement between the freezing characteristics and the soil moisture characteristic.

This idea was used to develop a laboratory device by Bittelli *et al.* (2003) [43], who named the device Thermo-Dielectric Analyzer (TDA), shown in Figure 9. The measurement principle is based on accurate measurements of temperature (by means of high accuracy platinum thermistors) and soil water content by Time Domain Reflectometry (TDR). The soil sample is placed into a sample holder where a TDR antenna was specifically designed for measuring soil water content. The sample temperature is progressively decreased by a Peltier cooler and simultaneous measurement of soil temperature and water content are performed at various temperatures below zero degree Celsius.

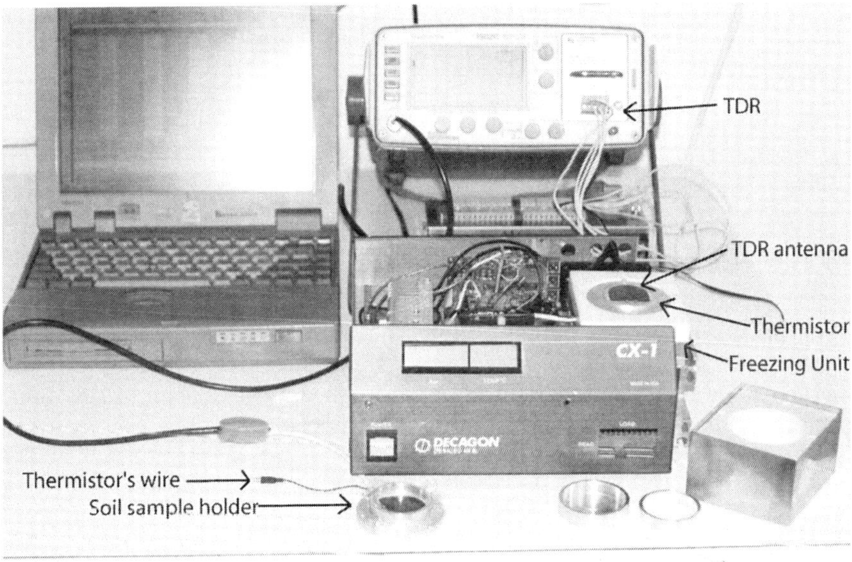

Figure 9. Thermo Dielectric Analyzer (TDA). Picture of the instrument with sample holder, TDR, TDR's antenna and Peltier's freezing unit. The Decagon CX-1 is only used as a case for holding the prototype developed by the author.

The method described here offered an alternative technique to measure the soil moisture characteristic of a porous media. The methods provided a complete soil moisture characteristic from −10 to −10-*5* J kg-*1*, covering a wide range of soil water potential. In fact, according to the Clapeyron equation, with a temperature measurement accuracy of 0.01 K, the water potential can be resolved to −11 J kg-*1*. The TDA is a fast (usually a complete cooling and melting cycle takes 12–15 hours) method to obtain a complete SWR curve. A disadvantage is that since water volumes changes with freezing the pore size distribution may be affected with changes in the soil water retention properties. However, according to the authors when compared to other methods, the effect was negligible.

Overall the advantages of: (a) provides a water retention curve spanning several orders of magnitude in water potential, (b) is fast. The disadvantages are: (a) it is not yet a commercial instrument, (b) usually only drying curves are obtained, (c) the measurement is affected by changes in soil water volume.

8. APPLICATIONS FOR SOIL WATER MANAGEMENT IN AGRICULTURE

In this section, a few examples about how measurements of SWP can be used in practical applications are provided.

8.1. Field monitoring

The first area focuses on continuous monitoring of SWP in the field for irrigation management. A common approach for quantifying the irrigation amount and scheduling is through the calculation of the plant available water (PAW), which is defined as the difference between the soil water content at field capacity minus the soil water content at the permanent wilting point. Measurement of field SWP allows to replenish the soil and to compute the necessary irrigation volumes. Irrigation management is commonly defined as choices of irrigation techniques, schedule and volumes. However, successful irrigation depends upon understanding and utilizing irrigation scheduling principles to develop a plan that is tailored to crop water requirements and to meteorological and soil conditions.

Irrigation scheduling methods should be based on: (a) soil and/or crop monitoring, and (b) soil water budget calculations. Knowledge of SWP is needed in both cases, in the first case for soil monitoring aimed to replenish the soil profile of the correct amount of water, in the second case to compute infiltration, water redistribution, deep percolation and evapotranspiration. As we mentioned earlier, the SWP is the driving force for water movement in the entire soil-plant-atmosphere system.

Figure 10 shows an example of SWP monitoring by using heat dissipation sensors. The data are unpublished data from the study presented by

Bittelli *et al.* (2006) [44]. The area is covered by a dense natural vegetation (various species of shrubs and herbaceous plants), typical of the natural settings of the Apennine Mountains. The soil is at, or close to the field capacity (FC) value (−33 J kg-*1*) for most of the winter at both depths. During the summer the SWP decreases to values lower than the wilting point (WP). Indeed, many of the spontaneous plants (since they were not irrigated) experienced water stress during the end of summer. At the end of summer, the SWP peaks are due to the intense rainfall events, also depicted by the precipitation graph.

While these data are representative of a natural area, under irrigated agriculture these measurements can be employed to schedule irrigation time and volumes, with the aim of replenishing the soil of water to bring the water potential values close or at field capacity. The irrigation system is controlled by a datalogger: when the SWP potential is recorded and the value is below a certain threshold, the datalogger switch on a port to start the irrigation pumps, when the SWP has reached a given value (for instance field capacity), the datalogger switch off the pumps. The system should be coupled with a meteorological station that records precipitation, temperature, relative humidity and wind velocity.

In many cases, techniques such as controlled water stress could also be employed to reduce irrigation water without reducing the crop yield to values that are not economically sustainable for the farm. Finally knowledge of soil and plants' water potential can be used to develop, calibrate and test crop growth models and water budget models

The description above provided an overview of the most common methods to measure the SWP. A more detailed description of these methods can be found on books on soil analysis [45]. However, how can these techniques be used to improve the water use efficiency in agriculture?

The answers can be grouped in two main areas: (a) measurement of soil hydraulic properties to be used for the quantification and modelling of the soil water budget and, (b) field installation of SWP sensor for irrigation management and scheduling.

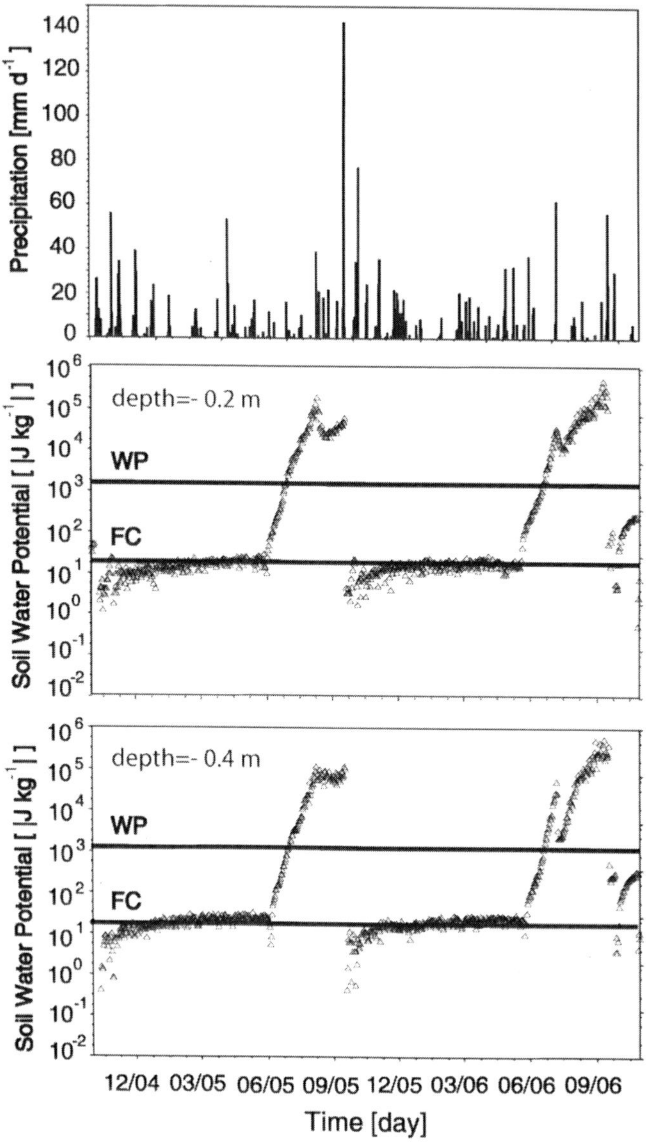

Figure 10. Daily soil water potential measured with heat dissipation sensors at two depths, under a natural spontaneous field of various species of shrubs and herbaceous plants, for a two years period. The graph on top is daily precipitation. The two horizontal lines indicate the value of field capacity (FC) and wilting point (WP).

8.2. Soil hydraulic properties, crop modelling and soil water budget

The second important application focuses on the necessity of obtaining a quantification of the soil hydraulic properties (*i.e.*, the soil water retention and the soil hydraulic conductivity), to be used to compute the soil water budget. There are two main approaches for obtaining the soil water retention parameters: the first is to experimentally determine the soil water retention and the soil hydraulic conductivity curve, and fit various functions to the experimental data [46,47,48,49]; the second is to derive the soil water retention parameters from knowledge of basic soil physical properties by using pedotransfer functions [50,51]. Pedotransfer functions (PTF) are functions obtained from different forms of regression analysis that allow to obtain the soil hydraulic properties from knowledge of basic soil physical properties such as texture, bulk density and soil organic matter content. While PTF's are very useful tools, for a better quantification of the soil water budget and modelling it is advisable to obtain experimental data of hydraulic properties therefore measuring the SWP combined with the water content (soil water retention curve) and the hydraulic conductivity (soil hydraulic conductivity curve).

Another application where the measurement of the SWP is necessary is to determine the value of field capacity (FC) and wilting point (WP), needed to compute the plant available water (PAW), which is given by PAW = WC*FC* – WC*WP*, where WC*FC* is the soil water content at field capacity and WC*WP* is the soil water content at the wilting point.

The FC can be defined as soil water content held in soil after excess water has drained away and the rate of downward movement has decreased. This value is usually reached within 2–3 days after a rain or irrigation has ceased. A quantitative description of FC is the soil water content retained in the soil at a SWP of −33 J kg-*1* for clayey and loamy soils, and at a SWP of −10 J kg-*1* for sandy soils.

The wilting point (WP) is defined as the minimal amount of soil water retained before the plant starts to permanently wilt. If the soil water content is further decreased below this value, the plant is unable to regain the cell turgor. A quantitative description is the soil water content retained at a SWP of −1,500 J kg-*1*. This value is considered valid for the most common agricultural crops (wheat, corn, sunflower), however plants adapted to dry environments are able to survive at lower SWP values.

Clearly, the measurement of the soil water content at specified potentials requires equilibration of the soil sample at specified SWP or the measurement of the SWP potentials at different water contents.

While the definitions of FC and WP are empirical and their value may change depending on soil type, crop and crop variety, for practical applications they are useful and commonly used for irrigation management and crop modelling. Indeed, many crop and irrigation models [52,53] use the value of PAW to estimate crop water requirements and manage irrigation scheduling. In these models, the value of actual plant transpiration is determined by functions depending on the PAW in the soil "bucket". Moreover, these models compute water flow based on the "tipping bucket" that utilizes the differences between FC and WP to determine the amount of water moving through soil layers in the vertical direction.

A correct measurement or estimation of the field capacity and wilting point volume fraction are therefore critical for obtaining reliable estimates of the above mentioned soil and crop related parameters, for land-use system analysis.

The computation of the soil hydraulic properties and related variables are then needed to compute the soil water budget. The soil water budget describes the changes in soil water content (within a given time interval), by computing the input of water (positive terms) and the output (negative terms) from a given volume of soil:

$$\Delta S = P + I - E - T - RO - DP - SLF \quad (5)$$

where ΔS is the change in soil water storage, P is precipitation, I is irrigation, E is soil evaporation, T is plant's transpiration, RO is surface runoff, DP is deep percolation and SLF is subsurface lateral flow. Often the evaporation and transpiration term are lumped into the term evapotranspiration. The SWP affects the magnitude of all the output terms E,T,RO,DP and SLF, therefore affecting the water budget as a whole. A correct quantification of the water budget, regardless whether or not it is computed through detailed crop modelling or through simplified formulations, should take into consideration this variable.

9. CONCLUSIONS

In this paper, the soil water potential has been first described to provide some definitions needed to understand its fundamental role in the soil water budget. The most important experimental techniques have been presented, with a detailed description of physical principles, most important technical features, advantages and disadvantages. Finally, some examples have been described with applications of SWP measurements in agriculture with the aim of saving irrigation water.

ACKNOWLEDGEMENTS

The use of trade and brand names in this report is for identification and scientific purposes only and does not constitute endorsement by the University of Bologna or the author.

REFERENCES

1. Hamdy, A.; Ragab, R.; Scarascia-Mugnozza, E. Coping with water scarcity: water saving and increasing water productivity. *Irrig. Drain.* **2003**, *1*, 3–20.
2. Dai, A.; Trenberth, K.E.; Quian, T. A global dataset of Palmer Drought Severity Index: Relationship with soil moisture and effects of surface warming. *J. Hydrometeorol.* **2004**, *5*, 1117–1130.
3. Pieri, L.; Bittelli, M.; Wu, J.Q.; Dun, S.; Flanagan, D.C.; Rossi Pisa, P.; Ventura, F.; Salvatorelli, F. Using the Water Erosion Prediction Project (WEPP) model to simulate field-observed runoff and erosion in the Apennines mountain range, Italy. *J. Hydrol.* **2007**, *336*, 84–97.
4. Timlin, D.; Pachepsky, Y.; Walthall, C.; Loechel, S. The use of a water budget model and yield maps to characterize water availability in a landscape. *Soil Till. Res.* **2001**, *58*, 219–231.
5. Bouman, B.A.M.; Tuong, T.P. Field water management to save water and increase its productivity in irrigated lowland rice. *Agr. Water Manage.* **2001**, *49*, 11–30.
6. Wang, D.; Kang, Y.; Wan, S. Effect of soil matric potential on tomato yield and water use under drip irrigation condition. *Agr. Water Manage.* **2007**, *87*, 180–186.
7. Thompson, R.B.; Gallardo, M.; Valdez, L.C.; Fernandez, M.D. Using plant water status to define threshold values for irrigation management of vegetable crops using soil moisture sensors. *Agr. Water Manage.* **2007**, *88*, 147–158.

8. Viswanatha, G.B.; Ramachandrappa, B.K.; Nanjappa, H.V. Soil-plant water status and yield of sweet corn (Zea mays L. cv. Saccharata) as influenced by drip irrigation and planting methods.*Agr. Water Manage.* **2002**, *55*, 85–91.

9. Irrinet system. Consorzio di bonifica per il Canale emiliano romagnolo (Cer). Available online: http://irrigation.altavia.eu/logincer.aspx (accessed on 10 February 2010).

10. Hillel, D. *Introduction to Soil Physics*; Academic Press: San Diego, CA, USA, 1982.

11. Jury, W.A.; Horton, R. *Soil Physics*; John Wiley and Sons: Hoboken, NJ, USA, 2004.

12. Richards, L.A. Porous plate apparatus for measuring moisture retention and transmission by soils. *Soil Sci.* **1948**, *66*, 105–110.

13. Richards, L.A. Physical condition of water in soil. In *Methods of Soil Analysis, Part 1. Agronomy Monograph*; Black, C.A., Ed.; American Society of Agronomy: Madison, WI, USA, 1965; pp. 128–152.

14. Campbell, G.S. *Soil Physics with Basic*, 1st ed.; Elsevier: Amsterdam, The Netherlands, 1985.

15. Campbell, G.S.; Norman, J.H. *An Introduction to Environmental Biophysics*, 2nd ed.; Springer-Verlag: New York, NY, USA, 1998.

16. Campbell, G.S.; Shiozawa, S. Prediction of hydraulic properties of soils using particle-size distribution and bulk density data. In *Indirect Methods for Estimating the Hydraulic Properties of Unsaturated Soils*; Van Genuchten, M.T., Leij, F.J., Lund, L.J., Eds.; University of California: Riverside, CA, USA, 1992; pp. 317–328.

17. Young, M.H. Piezometry. In *Methods of Soil Analysis, Part 4, Physical Methods*; Dane, J.H., Topp, G.C., Eds.; SSSA Book: Madison, WI, USA, 2002; pp. 547–571.

18. Sisson, J.B.; Gee, G.W.; Hubbell, J.M.; Bratton, W.L.; Ritter, J.C.; Ward, A.L.; Caldwell, T.G. Advances in tensiometry for longterm—monitoring of soil water pressures. *Vadose Zone J.* **2002**, *1*, 310–315.

19. *T5-X. User Manual*, Available online: http://www.ums-muc.de/en/products/tensiometer/t5.html(accessed on 10 November 2009).

20. Tamari, S.; Gaudu, J.C.; Simmoneau, T. Tensiometric measurement with metastable state of water under tensions. *Soil Sci.* **1993**, *156*, 149–155.

21. Hendrickx, J.M.H.; Nieber, J.L.; Siccama, P.D. Effect of tensiometer cup size on field soil water tension variability. *Soil Sci. Soc. Am. J.* **1994**, *58*, 309–315.

22. Flint, A.L.; Campbell, G.S.; Ellett, K.M.; Calissendorf, C. Calibration and temperature correction of heat dissipation matric potential sensors. *Soil Sci. Soc. Am. J.* **2002**, *66*, 1439–1445.

23. Or, D.; Wraith, J.M. A new TDR-based soil matric potential sensor. *Water Resourc. Res.* **1999**, *35*, 3399–3407.

24. MPS-1 Dielectric Water Potential Sensor Operator's Manual, Version 3.0. Decagon Devices: Pullman, WA, USA.

25. Saito, H. Monitoring variably-saturated water flow in volcanic ash soils using the matric potential sensor, MPS-1. In *Symposium: Spatial and Temporal Dynamics of Soil Water and Their Relations to Biotic and Abiotic Processes at Different Scales: II.*; ASA: Pittsburgh, PA, USA, 2009.

26. Wullschleger, S.D.; Dixon, M.A.; Oosterhuis, D.M. Field measurement of leaf water potential with a temperature-corrected *in situ* thermocouple psychrometer. *Plant Cell Environ.* **1988**, *11*, 199–203.

27. Campbell, G.S. Improved thermocouple psychrometers for measurements of soil water potential in a temperature gradient. *J. Phys. E: Instrumen.* **1979**, *12*, 113–121.

28. Spanner, D.C. The Peltier effect and its use in the measurement of suction pressure. *J. Exp. Bot.* **1951**, *2*, 145–168.

29. Richards, L.A.; Ogata, G. Thermocouple for vapor pressure measurement in biological and soil systems at high humidity. *Science* **1958**, *128*, 1089–1090.

30. Brown, R.W. *Measurement of Water Potential with Thermocouple Psychrometers: Construction and Applications*; USDA Forest Service Research Paper INT-80. United States Department of Agriculture: Washington, DC, USA, 1970.

31. Wiebe, H.H.; Brown, R.W.; Barker, J. Temperature gradient effects on in situ hygrometer measurements of soil water potential. *Agron. J.* **1977**, *69*, 933–939.

32. McAneney, K.J.; Tanner, C.B.; Gardner, W.R. An in-situ dewpoint hygrometer for soil water potential measurement. *Soil Sci. Soc. Am. J.* **1979**, *43*, 641–645.

33. Klute, A. Water retention: Laboratory methods. In *Methods of Soil Analysis. Part 1. Physical and Mineralogical Methods*, 2nd ed.; Klute, A., Ed.; American Society of Agronomy: Madison, WI, USA, 1986; pp. 635–662.

34. Stackman, W.P.; Valk, A.; Van der Harst, G. *Determination of Soil Moisture, Retention Curve. Sand Box Apparatus. Range pF 0 to pf 2.7*; Technical Report; Ed.; Eijkelkamp.: Giesbeek, The Netherlands, 1969; p. 20.

35. Scanlon, B.R.; Andrasky, B.J.; Bilskie, J. Miscellaneous methods for measuring matric or water potential. In *Methods of Soil Analysis. Part 4. Physical Methods*; Dane, J.H., Topp, G.C., Eds.; Soil Science Society of America: Madison, WI, USA, 2002; pp. 643–670.

36. Dane, J.H.; Hopmans, J.W. Hanging Water Column. Part 4. Physical methods. SSSA Book Ser. 5. In*Methods of Soil Analysis. Part 4. Physical Methods*; Dane, J.H., Topp, G.C., Eds.; Soil Science Society of America: Madison, WI, USA, 2002; pp. 680–683.

37. Romano, N.; Hopmans, J.W.; Dane, J.H. Suction Table. In *Methods of Soil Analysis. Part 4. Physical*; SSSA Book Ser. 5. Dane, J.H., Topp, G.C., Eds.; Soil Science Society of America: Madison, WI, USA, 2002.

38. Campbell, G.S. Soil water potential measurement: An overview. *Irrig. Sci.* **1988**, *9*, 265–273.

39. Gee, G.W.; Ward, A.L.; Zhang, Z.F.; Campbell, G.S.; Mathison, J. The influence of hydraulic nonequilibrium on pressure plate data. *Vadose Zone J.* **2002**, *1*, 172–178.

40. Bittelli, M.; Flury, M. Errors in water retention curves determined with pressure plates. *Soil Sci. Soc. Am. J.* **2009**, *73*, 1453–1460.

41. Schofield, R.K.; Botelho da Costa, J.V. The measurement of pF in soil by freezing-point. *J. Agric. Sci.* **1938**, *28*, 644–653.

42. Spaans, E.J.A.; Baker, J.M. The soil freezing characteristic: Its measurement and similarity to the soil moisture characteristic. *Soil Sci. Soc. Am. J.* **1996**, *60*, 1319.

43. Bittelli, M.; Flury, M.; Campbell, G.S. A Thermo-Dielectric Analyser to measure the freezing and moisture characteristic of porous media. *Water Resourc. Res.* **2003**, *39*, 1041.

44. Bittelli, M.; Salvatorelli, F.; Rossi Pisa, P. Correction of TDR-based soil water content measurements in conductive soils. *Geoderma* **2008**, *143*, 133–142.

45. Dane, J.H., Topp, G.C., Eds.; *Methods of Soil Analysis. Part 4. Physical Methods*; Soil Science Society of America: Madison, WI, USA, 2002.

46. Brooks, R.H.; Corey, A.T. Properties of porous media affecting fluid flow. *J. Irrig. Drainage Div.* **1966**, *72(IR2)*, 61–88.

47. Van Genuchten, M.T. A closed-form equation for predicting the hydraulic conductivity of unsaturated soils. *Soil Sci. Soc. Am. J.* **1980**, *44*, 892–898.

48. Vogel, T.; Cslerov, M. On the reliability of unsaturated hydraulic conductivity calculated from the moisture retention curve. *Trans. Porous Media* **1988**, *3*, 1–15.

49. Durner, W. Hydraulic conductivity estimation for soils with heterogeneous pore structure. *Water Resour. Res.* **1994**, *32*, 211–223.

50. Rawls, W.J.; Brakensiek, D.L.; Saxton, K.E. Estimation of soil water properties. *Trans. ASAE* **1982**,*25*, 1316–1328.

51. Schaap, M.G.; Leij, F.J.; van Genuchten, M.T. Rosetta: A computer program for estimating soil hydraulic parameters with hierarchical pedotransfer functions. *J. Hydrol.* **2001**, *251*, 163–176.

52. Marletto, V.; Ventura, F.; Fontana, G.; Tomei, F. Wheat growth simulation and yield prediction with seasonal forecasts and a numerical model. *Agric. For. Meteor.* **2007**, *147*, 71–79.

53. Stöckle, C.O.; Donatelli, M.; Nelson, R. Cropsyst, a cropping systems simulation model. *Eur. J. Agron.* **2003**, *18*, 289–307.

7

CHAPTER

Developing a Framework to Measure Watershed Sustainability by Using Hydrological/Water Quality Model

Aditya Sood[1*], William F. Ritter[2]

[1]Currently at International Water Management Institute (IWMI), Colombo, Sri Lanka

[2]Department of Bioresources Engineering, University of Delaware, Newark, USA

ABSTRACT

A framework is built, wherein hydrological/water quality model is used to measure watershed sustainability. For this framework, watershed sustainability has been defined and quantified by defining social, environmental and biodiversity indicators. By providing weightage to these indicators, a "River Basin Sustainability Index" is built. The watershed sustainability is then calculated based on the concepts of reliability, resilience and vulnerability. The framework is then applied to a case study, where, based on watershed management principles, four land use scenarios are created in GIS. The Soil and Water Assessment Tool (SWAT) is used as a hydrology/water quality model. Based on the results the land uses are ranked for sustainability and policy implications have been discussed. This results show that landuse (both type and location) impact watershed sustainability. The existing land use is weak in environmental sustainability. Also, riparian zones play a critical role in watershed sustainability, although beyond certain width their contribution is not significant.

Keywords: Sustainability, Development, Framework, Watershed, Resilience, Reliability, Vulnerability, Hydrological models application, Landuse

1. INTRODUCTION

The concept of sustainable development has been there since the mid twentieth century but it was promoted by UNEP in the 70s and later by the World Bank in the early 80s. Sustainability implies maintaining natural resources for future generations and sharing them in an equitable manner in the current generation [1]. In reality there is no clear consensus on how to achieve development that can be considered sustainable [2]. Some scholars try to define sustainability based on the level of substitution of human and non-human capital allowed [3,4]. Others have defined sustainability from urban perspective [5,6]. Wackernagel and Rees [7] look at sustainability from an ecological perspective. Vliet [8] suggests nine principles of sustainability whereas Afgan [9] has done a good job of compiling a set of definitions of sustainability from different organizations and perspectives. Goodland [3] broke the sustainability debate into three paradigms—technocentric paradigm, ecocentric paradigm, and sustaincentric paradigm. Another way to look at sustainability is by broadly classifying it as social, economic and environmental [10]. Each class of sustainability focuses on maintaining its own capital. From human society's perspective, natural capital can be considered as a "stock of environmentally provided assets" that provide useful service to humans [3]. From the ecological perspective, natural capital includes all the living and nonliving (like air, water, soil) matter. Out of the three classes of sustainability, environmental sustainability takes precedence [4]. Social and economic sustainability depends upon the success of environmental sustainability.

Many frameworks have been developed over the last two decades so as to try to measure the impact of human activity on the society and the environment. The approaches to develop these frameworks can be broadly divided into four groups: media approach, stress response approach, resource accounting approach, and ecological approach [11]. The media approach looks at the indicators from the environmental component perspective like air, water, land etc. The stress response approach considers the impact of human activity on the environment and the environment's response to that. The resource accounting approach deals with the whole life cycle of the project from the flow of natural resources perspective. Finally, the ecological approach also includes ecological models along with other models and monitoring techniques. Within these four groups, four major international models have been developed.

"Framework for the Development of Environmental Statistics (FDES) developed by the United Nations Statistical Office, relates environmental components to four information categories—social and economic activities, natural events; environmental impacts of activities/events; responses to environmental impacts; and inventories, stocks and background conditions. Pressure-State-Response (PSR) framework developed by the Organization for Economic Cooperation and Development (OECD), has three components that are measured by indicators—pressure that the human society puts on the environment; the resulting state or condition of the environment; and the response of the society to these conditions. It forms a closed loop since the response of the society further impacts the environment. The Driving Force-State-Response (DSR) framework developed by the Commission on Sustainable Development, groups the indicators in three categories—driving force indicators deal with human activities and processes that impact the environment; state indicators measure the state of sustainable development; and the response indicator looks at the policies and other responses to the state of environment and development. Agenda 21 is based on this framework. The Driving ForcePressure-State-Impact-Response (DPSIR) framework developed by the European Environmental Agency (EEA) and the Statistical Office of the European Communities (Eurostat) is a further enhancement of PSR and DSR framework. It breaks up indicators into five groups— driving force that deals with social, economic and demographic development; pressure measures the processes that have harmful effect on the environment; state indicators deal with the state of environment; impact indicators measure the impact of the state of the environment on society; and response is the reaction of various organizations and society.

To bring the debate of sustainable development from out of the theoretical realm and make it useful, sustainability indicators are required. These indicators are affected by the cultural, technical, economical, physical and social conditions [2]. Heintz [12] suggests four major categories of indicators—system capacities and their allocation, consequences of water allocation, effects on people and underlying processes and driving forces. Afgan [9] suggests grouping the indicators for each subsystem and then combining them into a sustainability indicator. Indicators should be large enough to be comprehensive but small enough to be practical and understandable [13]. Each region or watershed will have to come up with its own set of sustainability indicators but the underlying principle has to be the same. Barlow et al. [14] define water sustainability as the "development and use of water resources in a manner that can be maintained for an indefinite time without causing unacceptable environmental, economic, or social consequences". Gleick [15] defines sustainable water use as "the use of water that supports the ability of human society to endure and flourish into the indefinite future without undermining the integrity of the hydrological cycle or the ecological systems that depend on it".

Various attempts have been made previously to come up with set of indicators or an index to quantify measure and monitor the social and environmental impact of human development. Most of these indicators, like Human Development Index (HDI), Water Poverty Index (WPI), Climate Variability Index (CVI), Environmental Sustainability Index (ESI) have been developed to evaluate national trends at global level [16]. Efforts are being made at the national level, for e.g. in US, the Sustainable Water Resources Roundtable (SWRR) came up with over 400 indicators for water sustainability and SWRR selected 17 critical ones [13]. Among the recent indices development, Chaves and Alipaz, 2007 used the pressure-state-response and the concept of UNESCO's International Hydrology Program framework of hydrology, environment, life and policy issues (HELP) to define a watershed level watershed sustainability index (WSI). This study considers the previous five years for building and measuring their index. There are few studies targeted at regional or watershed level and none (to the best knowledge of the authors) that use hydrological models. Most of the studies done and indices developed have focused on the past performances. These studies, although helpful to measure the impact of existing policy implementations, do not provide framework to measure the impact of future policy implications. Also, these studies look at various nations and regions from the same lens. Chaves and Alipaz [16] study measures the percentage variation in multiple indicators in the watershed but ignores the variability of different watersheds. The framework presented in this research uses hydrological/water quality model at watershed scale to consider various policy implementation scenarios in terms of land use before they are actually implemented. It takes into account the unique characteristics, local demands of each watershed. For example, one watershed can have higher assimilation property than other and hence can withstand higher development pressure than another. This framework is an example of a progressive, bottom-up approach. Thus, this paper adds to the ongoing research in sustainable development.

2. METHODOLOGY AND DATA DESCRIPTION

For this research, the following steps were followed:

Step 1: River Basin Sustainability Index was defined and developed (discussed in Section 4);

Step 2: The calibrated and validated watershed model was developed (not covered in this paper);

Step 3: Land use scenarios were created by using GIS (discussed in Section 5.2);

Step 4: The land use of the calibrated model was replaced with each scenario and the model run for 50 years;

Step 5: The River Basin Sustainability Index was measured for each land use scenario;

Step 6: Resilience, Reliability and Vulnerability was measured for each land use scenario.

SWAT2000 version incorporated in the EPA's Better Assessment Science Integrating point and Nonpoint Sources (BASINS) (version 3.1) was used in the case study to model Millsboro Pond watershed in Inland Bays basin in southern Delaware region of USA. To analyze a watershed, the model divides it into sub-watersheds based on topography. The sub-watersheds are further divided into hydrologic response units (HRU) based on similar land use and soil properties. Major components of the model are hydrology, weather, erosion, soil temperature, crop growth, nutrients, pesticides, and agriculture management [17]. The water balance is the driving force behind all of model simulations. Hydrology simulation is based upon two divisions—the land phase of the hydrological cycle (which includes movement of water, nutrients, pesticides from precipitation till the main channel); and the water or routing phase of the hydrological cycle (which includes movement of water, nutrients and pesticides through the channel network to the watershed outlet) [18]. Surface runoff and infiltration are calculated from the modification of the Soil Conservation Service (SCS) curve number method or Green and Ampt infiltration method (but this method requires subdaily precipitation data). SWAT can estimate evapotranspiration by one of the three methods: Modified Penman Montieth, Hargreaves, and Priestley-Taylor. Sheet erosion and sediment yield are calculated based on the Modified Universal Soil Loss Equation (MUSLE). Channel sedimentation routing is estimated based on the modified version of Bagnolds sediment transport equation, which uses flow velocity to estimate transport concentration capacity. SWAT profiles the soil in multiple layers. Soil water processes include infiltration, runoff, evaporation, plant uptake, lateral flow, and percolation to lower layers. Nitrogen and phosphorus processes are handled in a similar manner as in the Erosion Productivity Impact Calculator (EPIC) model. Use of nitrogen and phosphorus by biomass from the soil is based upon supply and demand. Plant's nitrogen (N) demand is a function of total biomass and the N concentration in the biomass. Supply is based on N in the soil. Nitrogen in the water is calculated as a product of nitrogen concentration and volume of water. Dissolved phosphorus is calculated based on the concept of partitioning phosphorus into solution and sediment phases. It uses labile P concentration in the top 10 mm of soil, runoff volume and a phosphorus soil-partitioning coefficient. Stream nutrient dynamics uses the kinetic routines from the water quality model QUAL2E. SWAT also allows simulation of land management practices. Management practices are broadly divided into agriculture management, water management and urban areas. Some of the management practices in agriculture management include plant growth cycle, timing of fertilizer, type of tillage, pesticide application, and removal of plant biomass. The crop model is a simplification of the EPIC crop model. Water management includes irrigation, tile drainage, impounded/depresssional areas, water transfer, consumptive water use, and loading from point sources. For the urban areas, the model estimates quantity and quality of the runoff based upon the impervious cover that are either directly connected to the drainage system and those that are not directly connected.

The terrain elevation data was downloaded from the USGS in digital raster form as Digital Elevation Model (DEM). The DEM used in this research was 1-degree DEMs (3- by 3-arc second data spacing) provides coverage in 1- by 1-degree blocks. The surface water data in the form of reaches data was also downloaded from the National Hydrography Dataset (NHD) developed by USGS based on 1:100,000 scale data. The NHD supersedes USGS Digital Line Graph (DLG) hydrography data and the EPA Reach File Version 3 (RF3). The soil profile was created from the data downloaded from STATSGO developed by the National Cooperative Soil Survey. The weather data, which includes precipitation, maximum and minimum temperature, solar radiation, and wind speed, was obtained for the Georgetown station from the Office of Delaware State Climatologist's website. The groundwater level data was obtained from the Delaware Geological Society, University of Delaware's website. The hydrological data was downloaded from the National Water Information System of the USGS for the station 01484525 at Millsboro Pond outlet at Millsboro, DE. The nutrient data (i.e. nitrogen and phosphorus) for the Millsboro pond outlet was retrieved from the Delaware Inland Bays Water-Quality Database (DIBWQDB) distributed by the Delaware Geological Survey (DGS). This database was built as part of the "Nutrient Inputs as a Stressor and Net Nutrient Flux as an Indicator of Stress Response in Delawares' Inland Bays Ecosystem" (CISNet) and the "Inland Bays Tributary Total Maximum Daily Load" (IBTMDL) projects and has data from 1998 to 2002. For land use land cover (LULC), National Land Cover Database (NLCD) 2001 land cover was used. NLCD has been created through a cooperative project conducted by the Multi-Resolution Land Characteristics (MRLC) Consortium.

SWAT uses its own weather generator, which is based on the WXGEN weather generator model. The weather generator is used either to predict weather information in the future or fill in the gaps in the existing weather data. The accuracy of this weather generator was not evaluated in this study, but it has been widely used in multiple studies (e.g. most of the studies using SWAT model [19, 20]. Although there are shortcomings with this generator but this is still popular stochastic model that is available [21]. Additionally, the focus of this research is on the framework of sustainability and a more reliable weather generator, when available, can be used in future studies. For the weather generator to function, an input file containing the statistics of the historic weather needs to be input. It is recommended to use at least 20-years' worth of historic data. For this research, 50 years of historic data, i.e. from 1958 to 2007 was used. The historic weather data was collected from the same weather station (i.e. Georgetown station) that was used to calibrate and validate the model. A program called WXPM3020, developed and supported by the staff of SWAT development team, was used to generate the statistical values to be input in the SWAT weather generator model. Before using WXPM3020 program, the historic weather data was run through another program called WXGN3020, which simulates any missing data. The output from WXPM3020 was used as an input for the .wgn file in SWAT.

3. FRAMEWORK

Development in the context of this research means change in land use from its natural function to provide goods and services to the mankind. The change in land use can be either development for residential, commercial, recreation or agriculture purposes. Sustainability in a watershed context means development of the watershed so as to provide at the minimum the basic services to its stakeholders at all the times (including drought) without compromising the quality of water, its ability to perform its ecological function and reserving sufficient resources for the biodiversity to proliferate. Goods and services provided by the development is the economic benefit to the community. Basic services provided by water to humanity are safe water for drinking, water for sanitation, water for livelihood (either agriculture or manufacturing). The term "all the times" is significant for sustainability. Water, being a natural resource, is an uncertain commodity. Droughts and floods create stress in the watershed—both to humans and non-human population. The development of a watershed can be called sustainable only when it can "cope with and recover" from these stresses. Hence the goal of water sustainability is not to have enough water available all the time, but rather have the system in place so as to adapt to the changing water scenario in a region. The system should be resilient enough to recover from occasional drought or floods and hence be less vulnerable to the whims of nature. Also, from an ecological perspective, there is a minimum amount of water required to fulfill its ecological obligations especially for aquatic systems. During the extreme conditions, these obligations should be met to a level so as not to impact the biota adversely. Finally, since land use is the main management tool in the watershed management, it is necessary to consider the continuity of undeveloped areas in a watershed so as to provide an undisturbed haven to the local wildlife. Given enough space and resources, the local wildlife can prosper as long as there is no interference from the humans. Fragmentation of wildlife habitat can adversely interfere with biodiversity. Hence continuity of human-interference-free land use can lead to sustaining the biodiversity of the region (this is a narrower view of biodiversity—a broader view would include consideration at eco-region scale).

3.1. Reliability, Resilience and Vulnerability

Sustainability in a water system can be measured in terms of reliability, resilience and vulnerability [22]. It can be measured in terms of the time required by the system to return to the equilibrium state after it has been disturbed from the equilibrium (also called "engineering resilience" [23]). Klein et al. [24] consider vulnerability to be "degree of incapability to cope with" disturbances. While the resistance is the ability of the system to avoid disturbance, the resilience (as discussed above) is the system's capacity to respond. Fara [25] uses the definition of vulnerability as "the characteristics of a person or group in terms of their capacity to anticipate, cope with, resist and recover from the impact of a natural hazard".

In this research, the mathematical representation of reliability, resilience and vulnerability as developed by Loucks and Gladwell [22] have been used. Such methodology has been successfully used in engineering based water resource systems [22,26]. By defining the acceptable range for the indicator, it can be seen that how many times the system fails (i.e. indicators fall outside the pre-defined acceptable limits). Thus reliability has been defined as the probability of a system to be within the range of satisfactory values over a period of time under consideration. Resilience has been defined as a measure of the "speed of recovery" from an unsatisfactory condition. Finally vulnerability has been defined as the statistical measure of the extent of failure. These three parameters can be used together to measure sustainability in a quantitative way.

3.2. Indicators for Watershed Sustainability

Indicators for watershed sustainability can be divided into social development, environment and biodiversity (**Figure 1**). For this framework, it has been assumed that the economic system is constrained by the availability of the natural resources (land and water to support livelihoods). As long as these resources are used in a way to provide for livelihood, it will lead to economic wellbeing. Since the issues related to watershed sustainability are region specific, the indicators should measure the performance of the region relative to its requirements. Thus two regions with very different hydrological profile may be equally sustainable because the demands within the watershed may be different. The indicators are defined in **Table 1**. The range of all the indicators are from 0 (worst case) to 1 (best case).

3.3. River Basin Sustainability Index (RBSI)

As discussed earlier, river basin sustainability incorporates social, environmental and biological sustainability. Hence river basin sustainability index can be created by summing up the social, environmental and bio-diversity indicators as discussed above. Based on the region, some indicators could be more critical than others. Thus a proper weight needs to be applied to these indicators based on the subjective judgment regarding the issues for the region under study. This is shown in the following equation:

$I_{WS} = W_{WD}*I_{WD} + W_{WL}*I_{WL} + W_{RE}*I_{RE}$

$+ W_{WP}*I_{WP} + W_{WE}*I_{WE} + W_{UL}*I_{UL}$ where W_x is the weightage for "x".

The RBSI gives a measure of the watershed's sustainability on an annual basis. A natural system cannot be sustainable all the time. During extreme events like drought or flooding, a system becomes unsustainable. Applying the analogy of sustainable livelihoods to sustainable watershed, a good measure of watershed sustainability is the ability of the system to cope and eventually recover from the stresses of such extreme (non-sustainable) events.

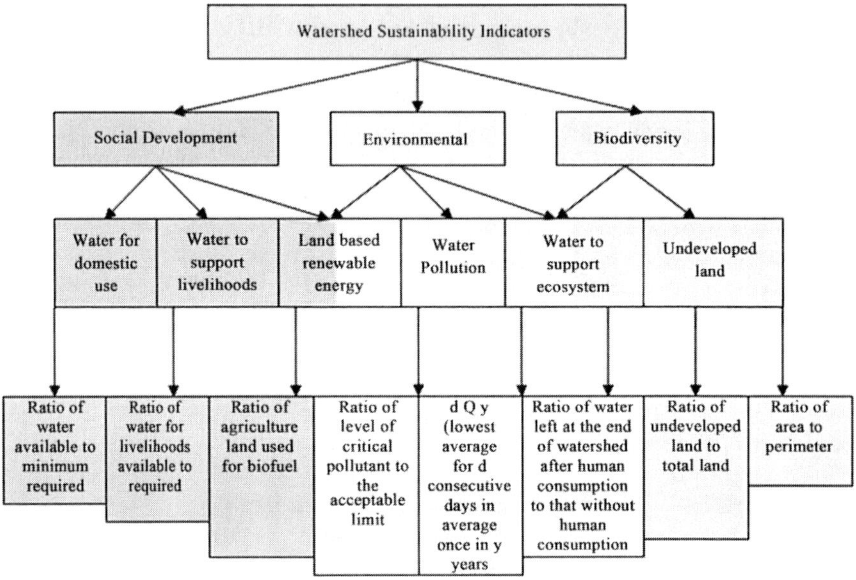

Figure 1. Watershed sustainability indicator.

Table 1. Descriptions of indicators used in this study.

Indicator	Time Step	Formula for Calculating Indicator	Comments
Water for Domestic Use (I_{WD})	Monthly	Water Available for Domestic Use/ Minimum Water Required for Domestic Use in theRegion	There are four basic water needs for humans: drinking water, water for hygiene, water for sanitation, and water for cooking. Based on the minimum requirements of a region, an index, which shows the capability of a watershed to meet that minimum requirement, is sufficient for this research.
Water to Support Livelihood (I_{WL})	Seasonal	Water Available for Livelihood Use/ Minimum Water Required for Livelihood Use in the Region	There is a strong relationship between sustainable livelihoods and water availability (Alan, 2000; Hope, 2003). From the watershed sustainability perspective, the indicator needs to measure the availability of water required to sustain the livelihood of the watershed. The water requirement will vary seasonally (as in case of water for irrigation), so the indicator can be calculated seasonally and then converted to the yearly average
Land Based Renewable Energy ((I_{RE})	Annual	Biofuel source produced within the watershed/Biofuel source required for the watershed population	Renewable energy is an important part of the equation of sustainable development. From the watershed sustainability perspective, some of the energy demands for the watershed should be met locally. From landuse perspective, renewable energy demand is related to renewable energy
Environment: Water Pollution (I_{WP})	Monthly	(Length of Stream that is not impared * No. of months it is not impared)/(Total Length of Stream in the Watershed * No. of Months Considered)	There are two sources of pollutants in water: point and non-point source. While point sources are easy to pinpoint , non-point pollution are hard.The indicator is calculated in monthly timesteps and looks at non-point pollution that is above the acceptable limits for that region
Water to Support the Ecosystem (I_{WE})	Monthly	Water Avaialable for the Ecosystem (i.e. water left after domestic and livelihood use)/Minimum Water Required	Various countries adopt different low-flow standards - A flow of frequency of 95% (Q95) , the seventh lowest average daily mean flow (MAM7) in a year, "the characteristic flow of low flow (DCE)", lowest 7-day average flow that occurs on average once every 10 years (7Q10)
Undeveloped Land (I_{UL})	Annual	(Land left undeveloped * Parimeter to Area Ratio)/(Minimum Land Required as Undeveloped for the Region * Acceptable Parimeter to Area Ratio)	The first step to preserve the biodiversity of a region is to preserve its undeveloped land. Thus a good indicator for this is the amount of land that is undeveloped with respect to the land in the region. But prevention of habitat loss alone is not the only requirement for preserving biodiversity. The other important factor is habitat fragmentation. The higher the percentage of undeveloped land with high area to edge ratio would imply better conditions for biodiversity.

3.4. Measuring Sustainability by Using Reliability, Resilience and Vulnerability Concepts

To measure such ability of a watershed, reliability, resilience and vulnerability can be measured by methodology developed by Loucks and Gladwell [22]. Once the RBSI is plotted (as shown in **Figure 2**), an acceptable range needs to be decided for the watershed. This decision is subjective based on the local conditions of the region. The acceptable range could have both the upper and lower limit or it could just have upper limit or only lower limit. For example if both drought and flooding is a critical issue in a watershed, the acceptable range would have lower and upper limit but if drought only is the main concern, then the framework could have only the lower limit, i.e. anything under that is unacceptable. The reliability, resilience and vulnerability are calculated as follows:

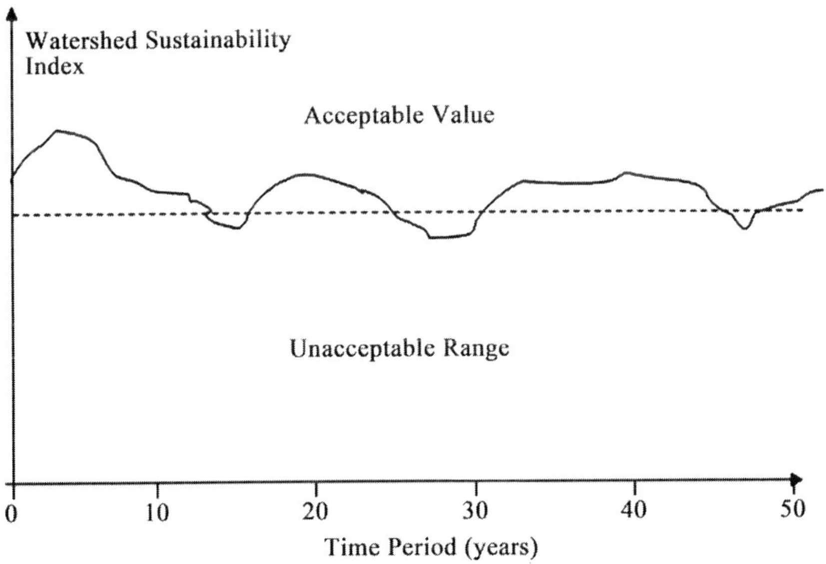

Figure 2. Watershed Sustainability Index (WSI) against time.

$$\text{Reliability}\left(S_{Rel}\right) = \frac{\text{Number of satisfactory values}}{\text{Total number of values}}$$

$$\text{Resilience}(S_{Res}) = \frac{\text{Number of Times a Satisfactory Value Follows an Unsatisfactory Value}}{\text{Number of Unsatisfactory Values}}$$

$$\text{Vulnerability}(S_{Vul}) = \text{Expected Extent given Unsatisfactory values} + \text{Expected Duration given Unsatisfactory values}$$

where

$$\text{Expected Extent given Unsatisfactory values} = \frac{\text{Cumulative extent of failure}}{\text{Number of individual failure events}}$$

and

$$\text{Expected Duration given Unsatisfactory values} = \frac{\text{Total number of failure periods}}{\text{Number of continuous series of failure events}}$$

Thus Watershed Sustainability

$$S_{Watershed} = S_{Rel} * S_{Res} * (1 - \text{Relative } S_{Vul})$$

where "Relative S_{Vul}" = Vulnerability/Maximum Vulnerability among all alternatives.

Relative S_{Vul} is subtracted from 1 to make it consistent with reliability and resilience.

4. CASE STUDY

4.1. Description of the Watershed

The Inland Bays is an estuary located in Sussex County in southeastern Delaware. It is one of the four drainage basins in the state of Delaware, draining into Atlantic Ocean. It is approximately 51 kilometers in size and drains roughly 810 square kilometers of watershed. The Millsboro Pond sub watershed is representative of other sub-watersheds in the Inland Bays basin. The Millsboro Pond watershed is 8708 hectare (87 Km^2) in area, out of which row crop agriculture occupies 41.8 percent, followed by 30.3 percent of deciduous forest cover and 11.5 percent of pasture land. The quality of the Inland Bays has been degraded over the years due to anthropogenic activities. The waters of the Inland Bays are rich in nutrients i.e. nitrates and phosphates. This nutrient enrichment can lead to eutrophication of the Bays. These conditions have been reported in the waters of Inland Bays as recently as 1998, 1999 and 2000 [27]. The nutrients entering the Bays are from various sources like agriculture, urban, wastewater and storm water. The watershed has also faced occasional droughts. The existing land use is highly fragmented. The current land use have about 45 percent of agriculture land, about 40 percent of forested land, roughly 9 percent of pasture land, 3 percent of water and wetland and rest is low residential development and other pervious land uses. This is a predominantly rural and agricultural watershed. According to the DNREC reports, 1256.7 kilograms (kg) of total nitrogen and 51.1 kg of total phosphorus are being added to the Bays daily.

4.2. Applying Framework to the Watershed

The prerequisite for applying the theoretical framework is to model a watershed as accurately as possible. The first step was to calibrate and validate the SWAT model. This was done as discussed in Sood and Ritter [28].

Two criteria that are critical to linking land use with water quantity/quality were considered to create land use scenarios. These are:

1) The recharge potential of the land: Andres [29], defines groundwater recharge as both the process and quantity of the precipitation that infiltrates through the unsaturated zone to the saturated zone.

2) The riparian zone along the water body: The vegetation in the riparian zones plays a critical role in the quality of water in the watershed.

Figure 3 shows the land use scenarios. The first (Case-1) is the existing land use (**Figure 3**(a)). The second scenario (Case-2) was created only on the basis of the recharge potential of the watershed (**Figure 3**(b)). Areas with high recharge potential should not only be left undeveloped, so that maximum water infiltrates, but it should also not be used for agriculture since it will cause high infiltration of nutrients (from fertilizers) into the groundwater system. For this research, a groundwater recharge map (and GIS data) provided by Delaware Geological Survey, University of Delaware [29] were used. The recharge potential of the watershed is classified as excellent, good, fair and poor. Thus keeping the percentage of types of land use the same, a hypothetical land use was created, with all the excellent, good, fair and poor recharge potential land being assigned to forest, pasture, agriculture, and development in that order. For the first land use scenario, no particular attention was paid to the riparian zones. Based on the second land use scenario, two more land use scenarios were created – one with 50 m riparian zone (Case-3) (**Figure 3**(c)) and the other with 200m riparian zone (Case-4) (**Figure 3**(d)). Changing proportions of land use type with in a watershed has economic implications, which is not being considered in this research.

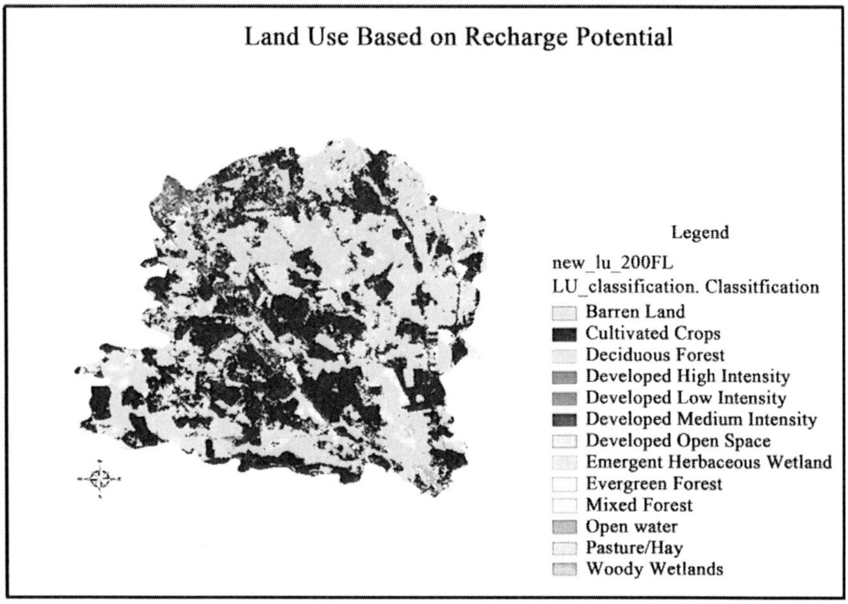
Land Use Based on Recharge Potential
Legend
new_lu_200FL
LU_classification. Classitfication
Barren Land
Cultivated Crops
Deciduous Forest
Developed High Intensity
Developed Low Intensity
Developed Medium Intensity
Developed Open Space
Emergent Herbaceous Wetland
Evergreen Forest
Mixed Forest
Open water
Pasture/Hay
Woody Wetlands

(a)

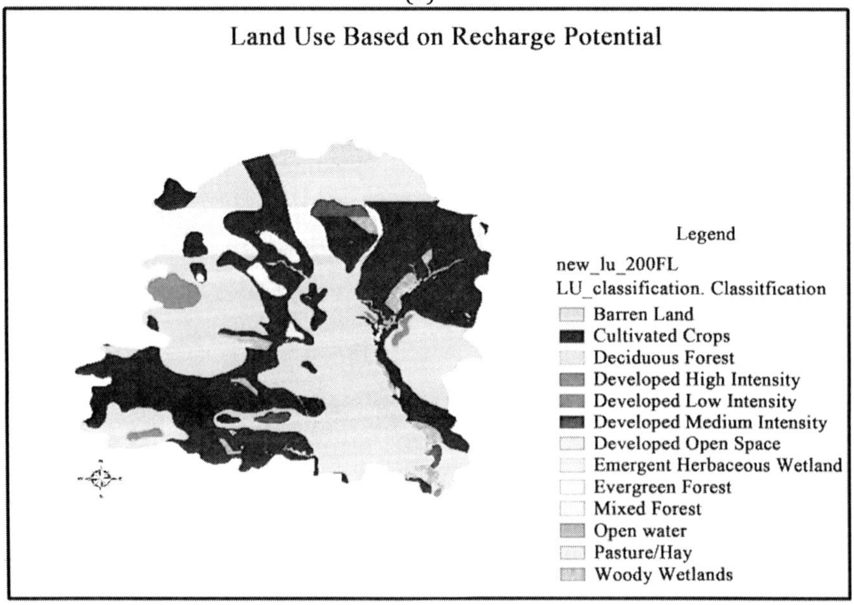
Land Use Based on Recharge Potential
Legend
new_lu_200FL
LU_classification. Classitfication
Barren Land
Cultivated Crops
Deciduous Forest
Developed High Intensity
Developed Low Intensity
Developed Medium Intensity
Developed Open Space
Emergent Herbaceous Wetland
Evergreen Forest
Mixed Forest
Open water
Pasture/Hay
Woody Wetlands

(b)

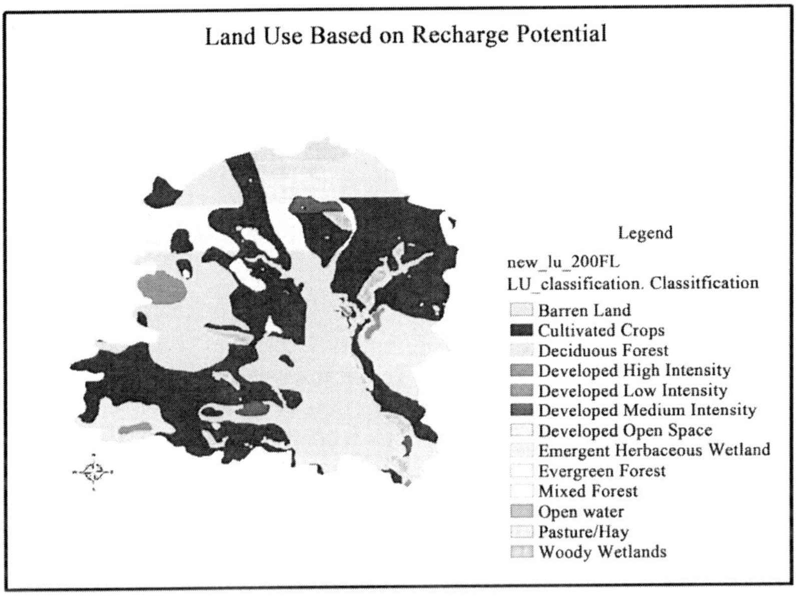

(c)

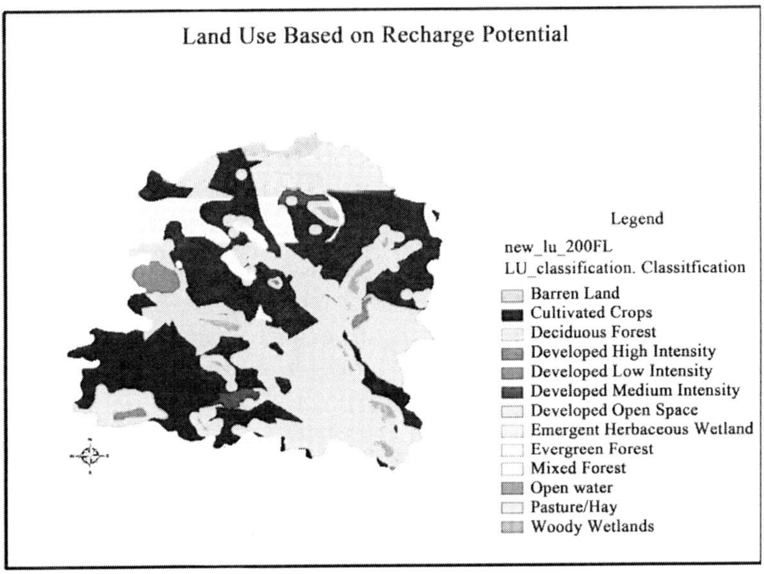

(d)

Figure 3. (a) Current land use—case 1. (b) land use based on recharge potential—case 2. (c) land use based on recharge potential and 50 m riparian zone—case 3. (d) land use based on recharge potential and 200 m riparian zone—case 4.

The calibrated SWAT model was the basis for the simulation of the new land use scenarios. The GIS data of the new scenarios were used to replace the old land use data. The model had to be reconfigured for each land use scenario. Because of this the calibrated values of the parameters were lost each time and had to be reassigned. The model was run with the different land use scenarios one at a time, with all the data collected and analyzed as discussed in the next section.

4.3. Indicators

The social, environmental and biodiversity indicators were calculated for 50 years. The total water available for human use and minimum stream flow was calculated as the sum of stream flow (which is made up of surface runoff and base flow) and deep aquifer percolation. It is safe to assume that the 100% of the percolation is available for human consumption because it is a renewable source and would not deplete the groundwater. The parameters used in calculating the indicators are shown in tables 2, 3 and 4.

After an extensive literature research, nothing substantive was found to suggest any specific combination of weightage. Weightage to environmental, social and biodiversity issues are region specific and is shaped by the priorities in a concerned region (or watershed) and how policymakers perceive the future threats. Since the priority of any region is availability of water, the direct use of water was given more weightage than other indicators. From a water use perspective, humans have been considered on par with other species. Thus the requirement for minimum flow for the ecosystem was given the same importance as water required for human consumption and livelihood. The greatest threat facing biodiversity is the dwindling undisturbed habitat. Pollution can be controlled by technological advances and energy demands of a watershed can be either managed better or energy can be imported from outside the watershed but there is no substitute for unaltered habitat. Thus biodiversity was given slightly more importance than pollution and energy requirements. In a scale of 100, domestic, livelihood and ecological water usage was given a weightage of 18 each, pollution and energy was given a weightage of 15 whereas biodiversity was given 16.

Table 2. Parameters used in calculating the hydrological indicator.

Parameter	Value	Unit
Amount used from percolation[1]	100	%
Area of watershed[2]	8704.1624	hectare
Population of Millsboro[3]	4000	
Population growth[4]	2	%
Domestic water consumption[5]	75	gallons/capita/day
Livelihood water consumption		
Amount of agriculture land[2]	45	%
Percent of agriculture land irrigated[6]	12	%
Agriculture land irrigated	470.02477	hectare
Water required for irrigation in the watershed[7]	236,632,038	gallons in June and August
Water required for chicken industry[8]	26	liter/broiler
Boilers produced in Delaware in 2007[9]	245,800,000	
No of flocks per year[10]	5.3	per year
Average flock size in Millsboro[11]	1,363,925	broilers/flock
Water consumed in poultry industry[12]	30	%
Water required for chicken industry[13]	4,000,000	gallons per season
Water required for ecological purpose[14]	9,505,161	gallons per day

[1]Percolation goes to deep aquifer. Since it is renewable, it can be assumed that 100% of the percolation can be made available for human consumption; [2]From SWAT Model; [3]Based on 2000 Census of population and housing, Delaware State Data Center; [4]Estimated based on the region; [5]Based on various sources - Handbook of Water Use and Conservation (Vickers 2001), NOAA, General Science Archive (http://www.newton.dep.anl.gov/askasci/gen01/gen01629.htm); [6]Based on Department of Agriculture; [7]Based on Department of Agriculture; [8]From the web site of United States Department of Agriculture, Agriculture Research Service; [9]From "FACTS ABOUT DELAWARE'S BROILER CHICKEN INDUSTRY" produced by Delmarva Poultry Industry, Inc., [10]from the web site of Spatial Analysis Lab, College of Agriculture and Natural Resources, University of Delaware—"Delaware Inland Bays Watershed Nutrient Management Project" document; [11]Calculated based on 9 and 10; [12]From Northcutt and Jones, 2004; [13]((Liter per broiler)*(No. of broilers per flock)*(No. of flock per year)/4)*(Percentage consumed)*(Area of Millsboro Pond/Area of Indian River catchment); [14]Using USGS's StreamStats Data-Collection Station Report—95 % duration flow.

Table 3. Parameters used in calculating the energy indicators.

Parameter	Value	Unit
Ethanol production	0.125	gal/Kg of dry corn
Bio diesel production	0.052	gal/Kg of wet soy
Wet/Dry soy ratio	1.3	
Energy per gallon of ethanol	80,000	BTU/gallon
Energy per gallon of bio diesel	140,000	BTU/gallon
Watershed population	4000	
Population growth	2	%
Energy consumption per capita per year	290	MBTU
Energy demand to be met by biofuels	3	%
Improvement in technology	2	%
Crop used for energy	50	%

Table 4. Parameters used in calculating the environmental (pollution) indicator.

Parameter	Value	Unit
TP loading rate	1.00	Kg/hectare/year
TN loading rate	22.00	Kg/hectare/year
TP loading rate	0.08	Kg/hectare/month
TN loading rate	1.83	Kg/hectare/month
Total Stream length	109.849	Km
Total watershed area	8704.16	hectare

For domestic indicator, it was assumed that the domestic water consumption in the region is 282 liters (75 gallons) per capita per day. The population of the watershed was taken as 4000 with a growth rate of 2%. Based on these parameters, the domestic indicator was calculated for each month and then averaged over the year. After the demand for domestic water needs was met, the remaining water was grouped together at three months intervals to measure the seasonal demand of water for livelihood. For the Millsboro watershed, there are two major livelihoods, i.e. agriculture and the poultry industry. About 12% of the agricultural land in the watershed is irrigated from June and August at the rate of roughly 5 cm per week. In the poultry industry, roughly 98.4 litters (26 gallons) of water are used per broiler [30]. Approximately 5.3 flocks per year are produced and each flock contains 1,364,000 broilers in the watershed. Based on this, the livelihood water demand for the watershed was calculated on seasonal basis and the indicator calculated. The minimum stream flow was based on the stream flow statistics collected by USGS at their gauging station at the Millsboro Pond outlet at Millsboro, DE (Station number 01484525). A 90 percentile flow of 0.76 m^3/sec was used as the minimum stream flow required from an ecological perspective.

For the energy indicator calculations, the per capita energy requirement of the population was taken as 306×10^9 joules per capita per year. The population in the watershed was taken as 4000 with a growth rate of 2%. It is assumed that only 3 percent of the total watershed energy demand will be met by bio fuels and 50% of biomass (i.e. corn and soybean) grown in the watershed will be used for energy. The yield (i.e. Kilograms per hectare) from the SWAT model was used as the biomass generated from the watershed. Also, it was assumed that due to technological innovation, the increase in production in the energy from the same amount of biomass will be at the rate of 2%. Since the land under agriculture was kept the same in all the scenarios, there was not much difference in the energy indicators for all the scenarios.

For the biodiversity indicator, GIS was used to calculate area and perimeter of the forested land in the watershed. The best case area to perimeter ratio was calculated by considering all the land area in the watershed as forested. This gave the value of the area to perimeter ratio as almost three times higher than the best case land use scenario.

5. RESULTS AND ANALYSIS

5.1. Indicators

Figures 4-9 show the values of all the six indicators over the period of 50 years (i.e. from 2009 to 2059). There are three years (i.e. 2013, 2046, and

2058), where the domestic water demands is not met. These years probably indicate drought situation. The livelihood indicator is similar in the most of the scenarios except in 2052 and 2053, when land use case 2 performs the worst among all the scenarios. The indicator dips frequently in the first decade and then once and then towards the end of the simulation period. Looking at the figure closely, it is evident that although none of them meet the ecological water demand all the time in the year, case 3 and case 4 of the land use perform better than other cases most of the time. From the perspective of nutrients in the water system, the current land use fares the worst. Although case 2 performs better than case 1 most of the times, it is not very different from the current land use performance. That is understandable as the case 2 lacks a good riparian zone. Numerous studies have shown that the riparian zone land use has a substantial effect on the nutrient pollution in the water system. Land use case 3 and 4 perform much better than the other two scenarios, although

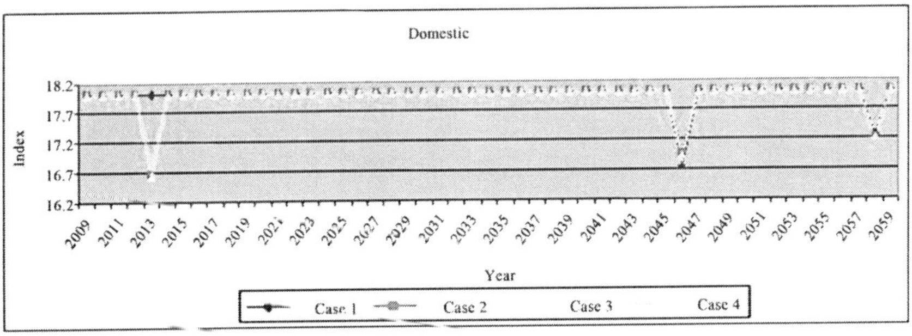

Figure 4. Plot of domestic indicator for the four land use cases.

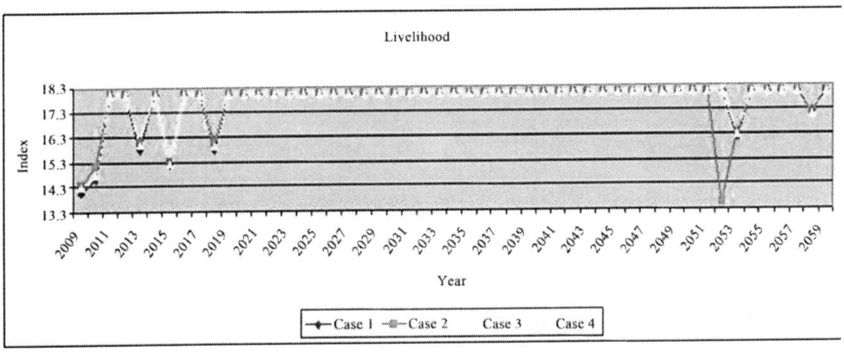

Figure 5. Plot of livelihood indicator over time (years) for the four cases of land use.

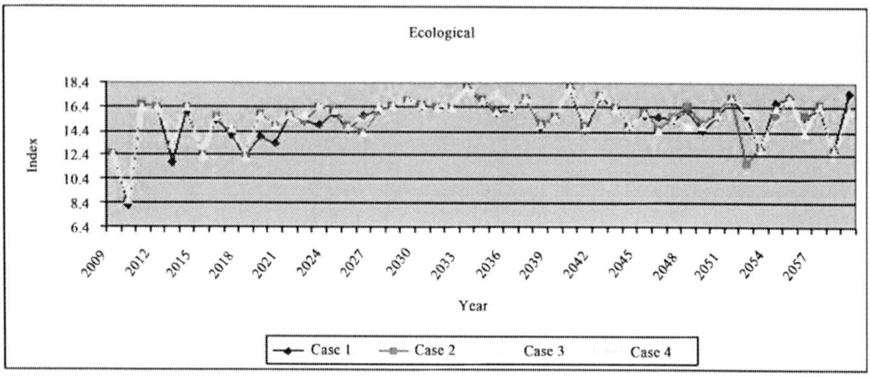

Figure 6. Plot of ecological indicator over time (years) for the four cases of land use.

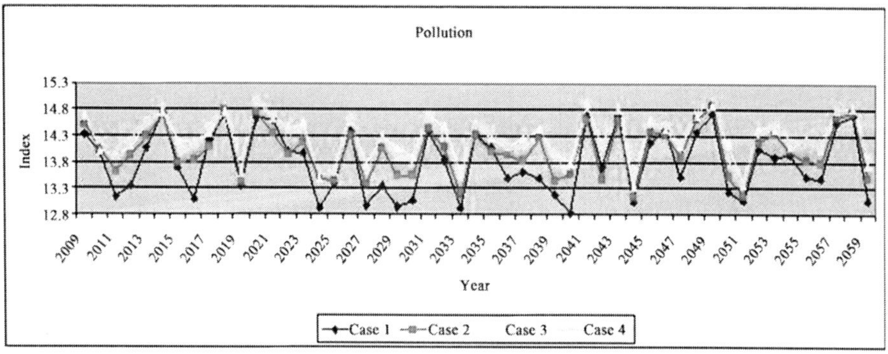

Figure 7. Plot of pollution indicator over time (years) for the four cases of land use.

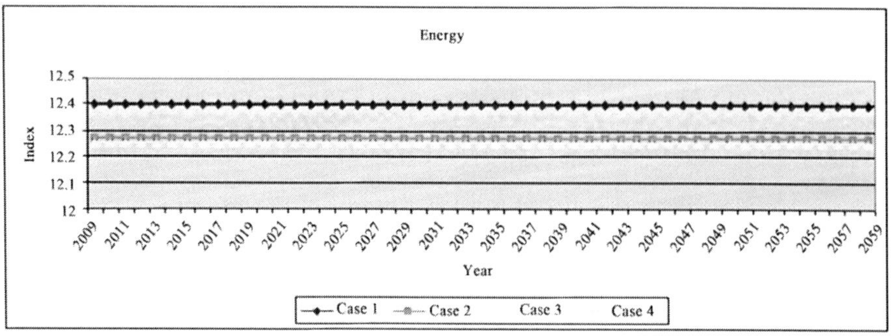

Figure 8. Plot of energy indicator over time (years) for the four cases of land use.

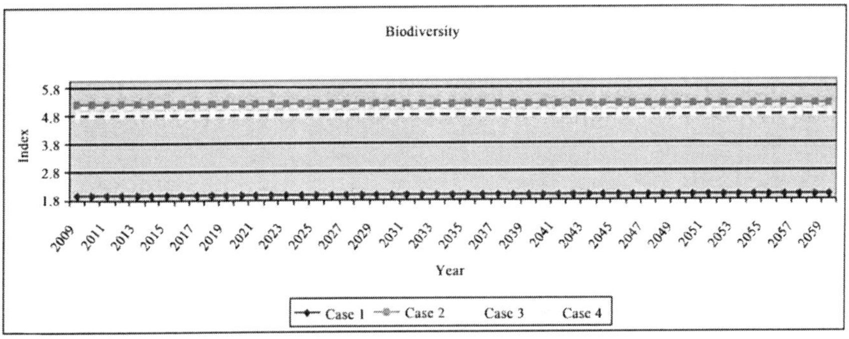

Figure 9. Plot of biodiversity indicator over time (years) for the four cases of land use.

the difference between them is not significant. This implies that there is not much difference in having a riparian zone 50 meter wide as compared to having a riparian zone 200 m wide for nutrient management. Thus it is clear that although a vegetative riparian zone in necessary to control the nutrients in the water way, but after a certain width, its effectiveness is reduced. The effectiveness of riparian zones in reducing nutrients is site specific, as it depends upon soil conditions, water table, type of vegetation etc. Thus to define an adequate riparian buffer width, each watershed needs to be modeled separately and appropriate riparian width defined. Land use cases 1 and 4 come up slightly higher than the other scenarios for energy indicator but as mentioned before, the difference is not significant. The current land use (case 1) is highly fragmented and as such had a very low area to perimeter ratio (roughly eight times less than the best case scenario). Among the four scenarios, case 2 has the highest value for the indicator.

5.2. Watershed Index

The watershed index was calculated by adding the weighted indicators for each of the land use scenarios. The indicators were then plotted against time (in years) for the next 50 years. The results are shown in **Figure 10**(a). On visual inspection, land use case 1 is much below the other scenarios and can be outright rejected. The majority of the difference is due to the biodiversity indicator. As discussed in the above section, the current land use, because of high fragmentation, fared badly with biodiversity indicator. That indicator outstrips (and hides) the significance of other indicators. Thus another graph was plotted without the biodiversity indicator (**Figure 10**(b)) and the watershed index compared. For the index without the biodiversity indicator, a different weightage, (Domestic—22, Livelihood—

22, Ecological—22, Pollution—17, Energy—17) was used. The new weightage was used to keep the ratio between the various indicators the same. In this case, it can be seen that "as is" scenario is comparable to other scenarios and in fact sometimes performs better that the case 2 (without any riparian zones).

5.3. Watershed Sustainability

To calculate the watershed sustainability, the first step is to define the acceptable value for the indicator. The decision on acceptable values for the indicator is subjective and is based on practicality and best judgment. Full availability of water for domestic, livelihood and ecological purposes was considered necessary. Hence they were all assigned a value of 1. On the other hand, if the pollution level falls to 0.8 of the "no pollution" value, could be considered acceptable. For energy, 70 percent of demand met would be considered acceptable. For the biodiversity indicator, 30 percent of the best possible scenario was considered because it is necessary to maintain the agriculture land to forest land ratio (since changing that would impact the economy of the region and it is not being considered in this research). Considering that 45 percent of the watershed land is agriculture, it will be hard to achieve a high value for this indicator. An unachievable value will make the whole system unsustainable and hide the other influence of other indicators.

Table 5 shows the results of the sustainability calculation for (a) all the indicators, (b) all indicators except the biodiversity indicator, and (c) all indicators (without biodiversity) and higher weightage to human use. In the first case, when all the indicators are considered, the current land use has the least reliability and resilience and very high vulnerability. This is because the current land use's biodiversity indicator is very low. Land use case 3 and 4 are almost similar, with case 4 just outstripping case 3. These two scenarios are less vulnerable and a little more resilient than case 2.

When the biodiversity indicator is removed, it is seen, that in fact land use case 2 is worse off than the current land use scenario. The vulnerability for the case 2 is higher. This can be explained due to the lack of the good riparian buffer in the case 2. Thus, although it has the highest biodiversity indicator value, its pollution indicator is lower than the other scenarios. Without the biodiversity indicator, land use case 3 is the most sustainable alternative of the four scenarios. Thus, it is seen that providing large riparian buffer widths does not necessarily make the system better off. If

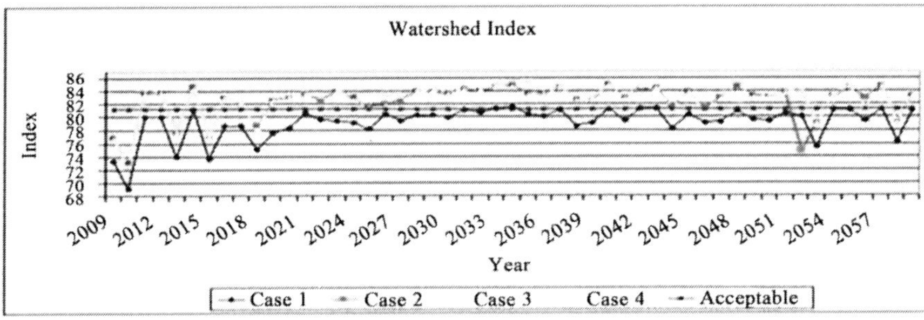

(a)

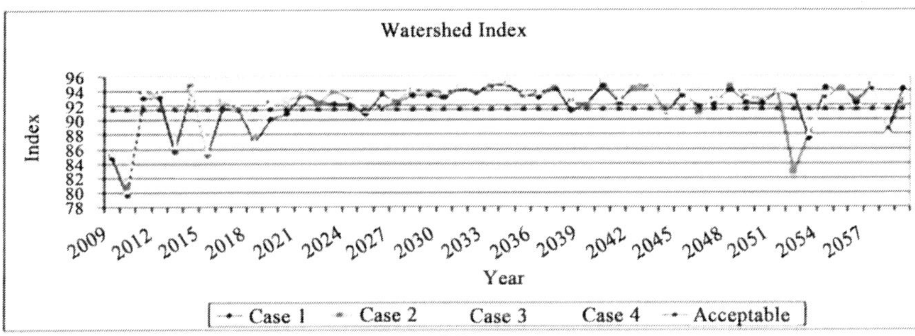

(b)

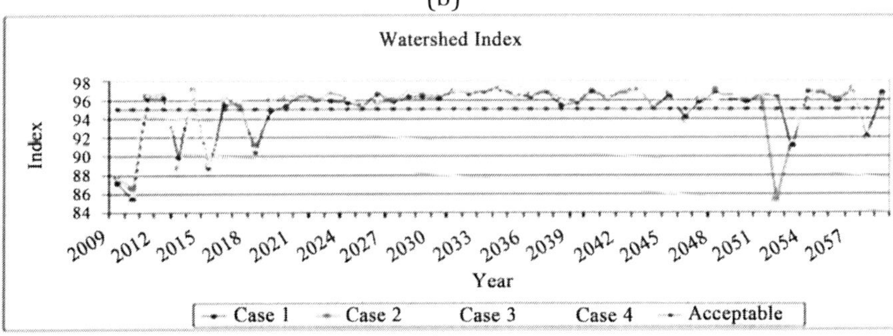

(c)

Figure 10. Watershed Index for the four land use scenarios plotted over 50 years (a) All indicators; (b) Without biodiversity indicator; (c) Without biodiversity indicator and weightage—Domestic—22, Livelihood—22, Ecological—22, Pollution—17, Energy—17.

biodiversity is ignored (i.e. zero weightage is given) and the ecological water needs, pollution and energy are given less importance (i.e. lesser weightage), the outcome is much different. The domestic water requirement and the livelihood water requirement were given a weightage of 35 whereas the other relevant indicators were given a weightage of 10. **Table 5**(c) shows the resilience, reliability, vulnerability and watershed sustainability for this new weightage scenario. In this case, the current land use scenario (case 1) is more sustainable than the other scenarios. Thus if biodiversity is ignored and other environmental and ecological issues are given less significance, the current land use works fine. In other words, the current land use is developed by keeping only anthropogenic requirements in mind and ignoring other ecological issues. Thus, it seems, the planning for existing land use, although meeting human needs, did not consider a holistic approach.

Table 5. Resilience, reliability, and vulnerability for (a) All indicators; (b) All indicators except biodiversity; (c) All indicators except biodiversity with more weightage to human requirements.

	(a) For all land use scenarios and biodiversity indicator included				(b) For all land use scenarios and biodiversity indicator excluded				(c) For all land use scenarios for the new weightage scenario (biodiversity indicator excluded)			
	Case 1	Case 2	Case 3	Case 4	Case 1	Case 2	Case 3	Case 4	Case 1	Case 2	Case 3	Case 4
Reliability	0.078431	0.843137	0.843137	0.843137	0.745098	0.764706	0.784314	0.803922	0.823529	0.823529	0.843137	0.843137
Resilience	0.042553	0.75	0.875	0.875	0.692308	0.75	0.909091	0.9	0.777778	0.777778	0.875	0.875
Extent Vulnerability	2.310879	4.24569	3.743589	3.420386	3.52429	4.133627	3.732527	3.997634	4.537963	5.410587	5.033877	5.0124
Duration Vulnerability	15.66667	1.333333	1.142857	1.142857	1.444444	1.333333	1.1	1.111111	1.285714	1.285714	1.142857	1.142857
Vulnerability	17.97755	5.579023	4.886446	4.563243	4.968735	5.466961	4.832527	5.108745	5.823677	6.696302	6.176734	6.155257
Relative Vulnerability	1	0.310333	0.271808	0.25383	0.908866	1	0.883951	0.934476	0.869686	1	0.92241	0.919202
Relative Watershed Sustainability	**0**	**0.436113**	**0.53722**	**0.550483**	**0.04701**	**0**	**0.082744**	**0.047408**	**0.083469**	**0**	**0.057242**	**0.059608**

6. DISCUSSION

The concept of reliability, resiliency and vulnerability has been applied previously to evaluate and select alternative design and operational policies for water resources projects [22,26,31,32]. But the scale of such an application has been restricted to a single project level. Also, there are not too many case studies using such a concept [26]. This research is an attempt to expand the scope of the concept of reliability, resiliency and vulnerability to the scale of watershed.

Most of the future global population growth is expected to take place in developing nations. Both, due to direct and indirect needs of increasing population and development, there is going to be extreme pressure on the limited land in these countries. This research shows that the location of development within a watershed is critical for sustainability of water resources and biodiversity. Along with how much land is converted to development, it is also necessary to consider where in the watershed that development is taking place. High recharge potential landuse should be either left in its natural state or used for the type of activities which require less impervious cover and have lesser scope of pollution infiltration. Also, riparian zones play a critical role in reducing the sediments and nutrient load in the streams flowing out of the watershed. Thus there should be adequate buffer between agriculture landuse and stream system. Based on the climatic conditions, this framework simulates the renewable water available over a long period of timeframe. It also takes into consideration minimum land and water required for biodiversity. Thus, in a sense, this framework helps in defining the carrying capacity of a watershed.

Although the framework proposed in this research was implemented in the watershed of a developed nation, it can be applied with slight modifications to the watersheds of the developing nations. In the developing nations the line between forested and developed land is a little blurred. There are subsistence communities that depend on non-timber forest products (NTFP) for their daily needs and livelihoods. A separate category of land use needs to be created. Many communities have water storage structures as "commons ponds", which have important significance in the social structure. The hydrological model needs to incorporate such ponds as they affect the water cycle in the watershed. These ponds would also play an important role in the supply side management of the water resources in the watershed.

The indicators to calculate the watershed sustainability could be created based on the critical issues of the watershed. Also the water needs and requirements in developing nations are different than those in the developed nations. So, although the indicators will stay the same, the acceptable values (denominators in the indicators used in this study) will change. Where excess water is an issue, for example in the floodplains and low lying areas, an upper limit, along with a lower limit needs to be considered for calculating reliability, resilience and vulnerability for the watershed. This framework could be used to identify the most effective adaptive measure, both in terms of its effectiveness and cost wise, to reduce the vulnerability of the watershed.

Some of the shortcomings of this framework are that the indicators used in this study are simplistic. Although the indicators cover the hydrological aspect well, they do not cover the social and biodiversity completely. Social issues and biodiversity are more complex than presented here and are outside the scope of this analysis. In a more detailed study, these aspects of landuse could be represented better by integrating other relevant models within this framework. Second, the land use scenarios in the case study were developed based on the recharge potential and riparian zones. Only four scenarios were created in this study. This limits the scope of land use possibilities. A more thorough study could include computer generated land uses linked to the model within the framework to generate the best land use after running hundreds of iterations. In deciding the land use scenarios, the computer program could use recharge potential, riparian zones, fragmentation in randomly creating land uses and/or any watershed specific parameter. Then the model could be run for these different land uses and a ranking given based on watershed sustainability calculated as defined in the framework. Then a subjective decision can be made by policy makers as to select the direction of future land use for the watershed. Third, although the SWAT model does a good job of hydrological analysis for surface runoff, root zone and channel flow, it does not include a detailed groundwater modeling that takes into consideration the coastal effect, large scale groundwater withdrawal and treated wastewater disposal using groundwater infiltration. Since groundwater is an important component of hydrological cycle and water use in the watershed, a dedicated model could be combined with the SWAT model for a better simulation and prediction of groundwater in the watershed. Forth, an important consideration in any watershed study is the scale of the watershed. A watershed can be as small as a catchment or as big as a river basin. Each scale would lead to a different outcome. The question is how to decide the appropriate scale of watershed to be considered for the study. If the watershed is too small, it will not be able to capture the holistic picture. If the watershed is too big, it would hide some of the local issues and not be as effective. Also a densely populated area like a big city will require much a bigger watershed to be sustainable within its boundaries. Thus watershed scale, which has not been an issue (because of low human population) in this study, needs to be incorporated in the framework.

Lastly, climate change has the potential to disrupt and modify hydrological regimes and thus affect watershed management. Along with land use scenarios, different climate change scenarios need to be considered. This research has not taken into consideration the future climate change scenarios, which has a potential to have a significant impact on the land use and water availability in the region.

7. CONCLUSIONS

This framework has many policy implications. Foremost, it provides a tool to policy makers, land use planners, zoning officials, watershed managers to look at land use in a holistic way. Currently, most of the planning is done using political boundaries. Thus land use is planned on the basis of local requirements within a district or a municipality boundary, without considering its impact on the larger "eco system" or the river basin. No consideration is given to population living outside the political boundary. In some instances, this leads to conflict between people. If the demand for the water is looked-upon at the watershed scale, such conflicts can be avoided. Since this framework takes into account the water requirement for the ecosystem, it also helps in reducing the conflicts with respect to water, between people and environment. The land use planning will become even more critical with the popularity of bio fuels and need for more land under cultivation. The greatest threat to biodiversity is the fragmentation of pristine land and interference due to human activity. Using recharge potential as a guiding principle for land use planning has two advantages—one, since the high recharge potential areas are left free of human development; it helps in recharging the groundwater faster. Also, infiltrating water is less polluted due to lack of human activities in these areas; second—since, recharge potential is based on soil properties and since soils with similar properties are grouped together, it is easier to plan with less fragmentation. Some of the limitations of this research, as mentioned in the section above, can be addressed in the future research. The biodiversity indicator needs to be further developed to better depict the biodiversity issues of a region. Also, an economic indicator can be added to the existing indicators to represent the social sustainability better. With the inclusion of an economic indicator, scenarios with less water intensive industries to replace agriculture (water intensive) land use could be played out and its affect analyzed with respect to sustainability, especially in the watersheds with high water scarcity.

REFERENCES

1. WCED (World Commission on Environment and Development), "Our Common Future," Oxford University Press, Oxford, 1987.
2. C. Tortajada, "Sustainable Development: A Critical Assessment of Past and Present Views," In: A. K. Biswas and C. Tortajada, Eds., Appraising Sustainable Development: Water Management and Environmental Challenges, Oxford University Press, Oxford, 2005.
3. R. Goodland, "The Concept of Environmental Sustainability," Annual Review of Ecology and Systemetics, Vol. 26, 1995, pp. 1-24.
4. R. Goodland and D. Herman, "Environmental Sustainability: Universal and Non-Negotiable," Ecological Applications, Vol. 6, no. 4, 1996, pp. 1002-1017.

5. D. Satterthwaite, "Sustainable Cities or Cities That Contribute to Sustainable Development?" Urban Studies, Vol. 34, no. 10, 1997, pp. 1667-1691.

6. G. McGranahan and D. Satterthwaite, "Urban Centers: An Assessment of Sustainability," Annual Review of Environment and Resources, Vol. 28, 2003, pp. 243-274.

7. M. Wackernagel and W. Rees, "Our Ecological Footprint, Reducing Human Impact on the Earth," New Society Publishers, Gabriola Island, 1996.

8. D. R. Van Vliet, "Sustainable Subdivision Planning and Design: Analysis, Literature Review and Annotated Bibliography," In Issues in Urban Sustainability: Institute of Urban Studies, University of Winnipeg, Winnipeg, 1994.

9. N. H. Afgan, "Sustainability Concept for Energy, Water and Environment Systems," In: Bogdan and Duic, Eds., Sustainable Energy Technologies, Swets and Zeitlinger B.V., Lisse, 2004.

10. R. Kranz, S. P. Gasteyer, H. Theodore, R. Shafer and A. Steinman, "Conceptual Foundations for the Sustainable Water Resources Roundtable. Universal Council on Water Resources," Water Resources Update, no. 127, 2004, pp. 11-19.

11. R. Shah, "International Frameworks of Environmental Statistics and Indicators," Inception Workshop on the Institutional Strengthening and Collection of Environment Statistics, 25-28 April 2000, Samarkand.

12. H. T. Heintz Jr., "Conceptual Foundations for the Sustainable Water Resources Roundtable," Water Resources Impact, Vol. 8, no. 4, 2006, pp. 7-10.

13. E. T. Smith and H. X. Zhang, "Overview/Introduction: Developing Indicators for the Sustainable Water Resources Roundtable," Water Resources Impact, Vol. 8, no. 4, 2006, pp. 3-6.

14. P. M. Barlow, W. M. Alley and D. N. Myers, "Hydrological Aspects of Water Sustainability and Their Relation to a National Assessment of Water Availability and Use," Water Resource Update, Vol. 127, No. 1, 2004, pp. 76-86.

15. P. H. Gleick, "Water in Crisis: Paths to Sustainable Water Use," Ecological Applications, Vol. 8, no. 3, 1998, pp. 571-579.

16. H. M. L. Chaves and S. Alipaz, "An Integrated Indicator for Basin Hydrology, Environment, Life, and Policy: The Watershed

Sustainability Index," Water Resources Management, Vol. 21, No. 5, 2007, pp. 883-895.

17. C. Santhi, R. Srinivasan, J. G. Arnold and J. R. Williams, "A Modeling Approach to Evaluate the Impacts of Water Quality Management Plans Implemented in a Watershed in Texas," Environmental Modelling & Software, Vol. 21, No. 8, 2006, pp. 1141-1157.

18. S. L. Neitsch, J. G. Arnold, J. R. Kiniry, J. R. Williams and K. W. King, "Soil and Water Assessment Tool Theoretical Documentation, Version 2000," Grassland, Soil and Water Research Laboratory, Blackland Research Center, Temple, 2002.

19. P. W. Gassman, M. R. Reyes, C. H. Green and J. G. Arnold, "The Soil and Water Assessment Tool: Historical Development, Applications, and Future Research Directions," American Society of Agricultural and Biological Engineers, Vol. 50, No. 4, pp. 1211-1250.

20. R. D. Harmel, C. W. Richardson and K. W. King, "Hydrologic Response of a Small Watershed Model to Generated Precipitation," Transactions of the ASABE, Vol. 43, No. 6, 2000, pp. 1483-1488.

21. T. W. R. Wallis and J. F. Griffiths, "An Assessment of the Weather Generator (WXGEN) Used in the Erosion/ Productivity Impact Calculator," Agricultural and Forest Meteorology, Vol. 73, No. 1-2, 1995, pp. 115-133.

22. P. D. Loucks and J. S. Gladwell, "International Hydrology Series—Sustainability Criteria for Water Resource Systems," Cambridge University Press, Cambridge, 1999.

23. L. H. Gunderson, "Ecological Resilience—In Theory and Application," Annual Review of Ecology and Systematics, Vol. 31, 2000, pp. 425-439.

24. R. J. T. Klein, M. J. Smit, H. Goosen and C. H. Hulsbergen, "Resilience and Vulnerability: Coastal Dynamics or Dutch Dikes?" The Geographical Journal, Vol. 164, No. 3, 1998, pp. 259-268.

25. K. Fara, "How Natural Are 'Natural Disasters'? Vulnerability to Drought of Communal Farmers in Southern Namibia," Risk Management, Vol. 3, No. 3. 2001, pp. 47-63.

26. T. R. Kjeldsen and D. Rosbjerg, "Assessment of the Sustainability of a Water Resources System Expansion," Integrated Water Resources Management, Vol. 272, 2001, pp. 151-156.

27. DNREC, "Inland Bays/Atlantic Ocean Basin," Delaware Department of Natural Resources and Environmental Control, Dover, 2001.

28. A. Sood and W. F. Ritter, "Evaluation of Best Management Practices in Millsboro Pond Watershed Using Soil and Water Assessment Tool (SWAT) Model," Journal of Water Resource and Protection, Vol. 2, No. 5, 2010, pp. 403-412.

29. A. S. Andres, "Ground-Water Recharge Potential Mapping in Kent and Sussex Counties, Delaware," Delaware Geological Survey, Report of Investigations No. 66, University of Delaware, Newark, 2004.

30. J. K. Northcutt and D. R. Jones, "A Survey of Water Use and Common Industry Practices in Commercial Broiler Processing Facilities," Journal of Applied Poultry Research, Vol. 13, No. 1, 2004, pp. 48-54.

31. T. Hashimoto, J. R. Stedinger and D. P. Loucks, "Reliability, Resiliency, and Vulnerability Criteria for Water Resource System Performance Evaluation," Water Resources Research, Vol. 18, No. 1, 1982, pp. 14-20.

32. S. K. Jain and P. K. Bhunya, "Reliability, Resilience and Vulnerability of a Multipurpose Storage Reservoir," Hydrological Sciences Journal, Vol. 53, No. 2, 2008, pp. 434-447.

8

CHAPTER

Using an Integrated Participatory Modeling Approach to Assess Water Management Options and Support Community Conversations on Maui

Andrea M. Bassi [1,*], John Harrisson [2] and Rushil S. Mistry [1,3]

[1]*Millennium Institute, 2111 Wilson Blvd, Suite 700, Arlington, VA 22201, USA*

[2]*Maui Economic Development Board, 1305 N. Holopono St, Kihei, HI 96753, USA*

[3]*McGill University, 845 Sherbrooke St. W. Montreal, Quebec, H3A 2T5, Canada*

ABSTRACT

The purpose of this study is to provide an integrated analysis of water distribution on Maui and the cross-sectoral impacts of policies and regulations aimed at rejuvenating and sustaining the deep-rooted culture on the island. Since the water diversion system was implemented in 1876 on the island of Maui, there has been contention among local interest groups over the right way to manage and allocate this precious resource. There is also concern over the availability of the precious resource in the long term, as the demand for water is expected to exceed the potential supply of water on Maui by 2020. This paper analyzes various long run scenarios of policy options presently being discussed on Maui. By collaborating with local experts, business leaders, and community members, to develop a tool that facilitates policy formulation and evaluation, informed decisions can then be made by the local community to ensure sustainable development.

Keywords: Maui; water supply; integrated modeling; system dynamics; water treatment; water management; sustainability; community development; strategies and policies

1. INTRODUCTION

Founded in 1905, Maui County consists of the islands of Maui, Kahoolawe, Lanai, Molokai and Molokini. With a total *de facto* population of 185,000 people in 2008 [1], the County of Maui, the third largest in the State of Hawaii, is projected to accommodate over 260,000 people by 2030 [2]. Among others, concerns have been expressed about the availability of water and energy to sustain such growth, as well as about the vulnerability of Maui's economy to energy and food imports and price volatility.

Historically, the traditional economy of Maui was based on agriculture, which provided the original inhabitants the necessary food staples—but with the arrival of American and European immigrants in the late 18th century, the sugarcane and pineapple industries were born and agriculture became an integral element of Maui's economy, occupying more than half of the land within the County [2]. Acting as a driving force for local development over the last century, the agriculture sector that Maui once had is much diminished. With the total value of crop sales in 2004 approaching $125 million and providing 1,850 jobs, agriculture only accounts for less than 2% of the total gross domestic product of the County [1]. Following the decline in the agriculture sector, tourism became the fundamental driver of the county's economy, accounting for 40% of economic activity in Maui County today [2]. With brief interruptions, Maui County's tourism industry has shown moderate growth over the years at about 1.5% annually, with average daily visitors amounting to approximately 52,000 people [3].

High population and economic growth on Maui, coupled with declining precipitation, have led to an increase in overall demand for water. Being the most fundamental aspect of all biological systems, water plays an integral role in every ecological process and has become a contentious issue on Maui. Due to a steep geological gradient present on the island, parts of Maui are endowed with large amounts of water, whereas other areas suffer from low precipitation and long periods of drought conditions [4]. In fact, being the second largest island in the Hawaiian Archipelago, two large shield volcanoes, the West Maui Volcano (Mauna Kahalawai or West Maui Mountains) and the East Maui Volcano (Haleakalā), dominate the landscape and heavily influence the local climate [5]. More specifically, due to the orographic lift that occurs when an air mass is forced from a low elevation at the base of the volcanoes to a high elevation at the summit, a rain shadow is created and drastic changes in climate occur. With moisture brought in by the prevailing Trade Winds from the northeast, the windward side of the island receives a significant amount of rainfall, whereas the leeward side experiences mostly arid conditions [4].

Connected by a 5-mile wide isthmus covered in terrestrial and marine sedimentary deposits, numerous streams and river systems have been carved into these geological giants over time [5]. Included in these streams are the Waikapū, ʻĪao, Kahakuloa, Honokōhau, and the Waiheʻe Rivers. Originating near the summit of Mauna Kahalawai, these rivers are fed by a plethora of rainfall received each year, averaging almost 350 in./yr (or 8,890 mm/yr). Conversely, the desiccated area surrounding Haleakalä experiences far lower precipitation, which accounts for the staggering gradients present on the island. With mean annual rainfall generally less than 100 in./yr (or 2,500 mm), parts of the island often enter into prolonged drought conditions during the dry seasons [6].

To allow for the development of the sugarcane industry throughout the island, complex diversion systems were put in place to redirect water from the rainy windward slopes of Haleakalä to the dry sugarcane fields in Central Maui. In 1876, Samuel Alexander and H.P Baldwin, the founders of Alexander & Baldwin, were partners in an expanding, but struggling sugar company. To solve the severe water problem that threatened to put them out of business, the company built a 17-mile irrigation system that diverted up to 60 Mgal/day [7]. Along with Alexander & Baldwin, other companies such as HC&S, founded in 1882, began developing ditch, tunnel and reservoir systems to redirect water in order to satisfy a growing demand on the leeward side of the island [8]. Today, almost 274 Mgal/day of surface water is diverted each year, with almost half of that water originating from the eastern part of the island [9].

The ʻĪao Aquifer System is the principal source of domestic water supply on Maui [9]. Withdrawals from this aquifer system totaled almost 18.94 million gallons per day (Mgal/day) during 2005; a significant increase from the 10 Mgal/day extracted during 1970 [10]. The Hawaii State Commission on Water Resource Management (CWRM) has determined the sustainable yield for this aquifer is 20 Mgal/day, which means it is operating at 95% of its sustainable yield [11]. The CWRM also indicates that the Kahului Aquifer System within the Central Aquifer System is being over-pumped by over 2,500%; along with the entire Central Aquifer System currently operating 168% over its sustainable yield [11].

With diminishing levels of supply and increased economic growth, future projections for this precious resource show decreasing levels of supply and increased levels of demand, where demand could exceed supply before 2020 [2,3,11,12]. The total available supply of water for the Central Maui System is estimated to be around 25 Mgal/day whereas the average daily demand for the system in 2005 was 21.39 Mgal/day. With projections indicating that demand for resources from the Central Maui System will reach 34.11 Mgal/day–9.11 Mgal/day over the current supply, by 2030, concern over the status of the future availability of water is mounting [2].

The water system on Maui also holds a strong cultural relevance and value to the local population. The practice of cultivating taro—a traditional staple in the indigenous diet and a culturally significant crop, relies heavily on the availability of fresh water for optimal growing conditions [13]. Growing taro is an integral element to sustaining Hawaiian traditions and it holds cultural significance and value to its people. As water resources are being diverted from the streams that provide water for taro cultivation, less water is available for taro production, and the crops are more susceptible to malnutrition, disease and other ailments. The local (mostly Native Hawaiian) community is taking a stance on these issues and taking legal steps to restore the natural levels of the streams to rejuvenate and sustain the local culture and traditions. In addition, the river system provides a home and habitat for a diverse range of species—such as native freshwater shrimp, small fish, mollusks and insects—and is part of a delicate balance of the unique and fragile ecosystem that has existed for thousands of years.

In an effort to inform decision-making on water resources and to plan for sustainable growth on the island, the Millennium Institute (MI) entered into a cooperative agreement with the Maui Economic Development Board (MEDB) to improve overall understanding of the water system on Maui and to provide the community with a tool to inform decision-making and support policy formulation and evaluation.

This paper provides a general overview of how an integrated approach utilizing Millennium Institute's Primary Country Model [14], renamed Primary Community Model (PCM) for this exercise, and fully customized to Maui, was applied to a system-wide analysis of water resources in Maui County. The intent of the analysis was to demonstrate the efficacy of implementing an integrated approach to modeling—utilizing PCM, to provide useful decision support services for water allocation and preservation employing "a broader approach to the consequences than implied by narrow output measures" [15].

2. LITERATURE REVIEW

A variety of models, each with their own unique attributes, investigate integrated approaches towards understanding both the biophysical and socio-economic impacts of water allocation [16]. In most cases, optimization models are used to assess water management issues and provide policy insights. These models, which are normative and prescriptive by nature, are used to provide a statement of the most favorable way to achieve a goal [17], that is, they provide a snapshot of what should be done today, normally in the most economical way, to optimize the current situation to reach a future goal. Optimization does not clearly indicate a strategy for action and does not provide insights on when the goal could be reached.

More specifically, Dinar *et al.* and Griffin *et al.* are examples of integrated optimization models applied to water management [18,19]. Dinar *et al.* attempt to estimate agricultural water demand and supply as a series of optimization functions, whereas Griffin *et al.* assign a value to in-stream flows and then develops optimization algorithms by varying economic policies and water consumption to assess the impact on water markets. Both of these approaches attempt to solve the water allocation problem from the demand component of the integrated model [18].

These models provide information on what to do to make the best of a given situation, but do not generate insights on what might happen in such a situation or what the impact of certain actions may be [21] outside of the water sector. The optimization models analyzed in this study can be best described as vertically integrated, representing detailed complexity coherently and in depth. They analyze the chain of physical steps for water from source to end use within a system and investigate the best ways to bring water to its customers, minimizing the cost of supply. As effective as these approaches may be, they are limited in their representation of the interaction between biophysical and socio-economic components within a given system.

To move from this state of vertical integration towards a more horizontal, flexible and comprehensive method of analysis, dynamic complexity must be accounted for. An example of this is the AQUA model presented by Hoekstra—the model that this study uses as its foundation for the water supply sector. Developed as a generic framework to be applied at different spatial scale levels, the aim of the model is to include as many of the interrelationships and feedback mechanisms between water, the environment and human beings as possible [21]. By employing a Pressure-State-Impact-Response (PSIR) ordering mechanism, a variety of unique interactions and relationships can be expressed within the model.

However, the AQUA model is narrow in scope, lacking connections to the energy sector, society as a whole, economic activity and the surrounding environment. Similar problems characterize System Dynamics (SD) model, that despite being dynamic are often too general or too narrow for being particularly relevant to this analysis (for an extensive review of the evolution of SD water-focused models see Winz and Brierley [22]). One of the major innovations introduced by the Millennium Institute and the Focus Maui Nui report prepared by MEDB [23], is to include the context of Maui County (*i.e.*, social, economic and environmental factors) into the integrated analysis of water. MI and MEDB therefore collaborated to create a comprehensive quantitative simulation model that can not only provide insights on water availability and allocation, but also contribute to the analysis of the interrelationships it has with other sectors [20].

In this research work, by creating a customized water sector for Maui; developed in collaboration with a group of local water experts, many characteristics of real systems, such as feedback loops, non-linearity and delays are incorporated in the water analysis that is carried out. These are based on the identification of the casual relationships that underlie the system analyzed—both within the water sector and across social, economic and environmental spheres. The modeling exercise focuses on the longer term impacts of policy choices and presents an aggregated picture of the social, economic and environmental development of Maui, filling a gap represented by a variety of very detailed sectoral studies (e.g., USGS, Water Resource Protection Plan, *etc.*).

Moreover, the model is also developed at the community level as well. Through collaborations with local experts, government officials, community members and businesses, the model is modified and customized the meet the diverse needs and specifications of the community at large.

3. RESEARCH APPROACH

The integrated analysis of the water sector in Maui County involved developing social, economic and environmental profiles of the islands, including the collection and processing of historical data, and the construction of a PCM for Maui, supported by open sessions during which experts and community members contributed to the building and customization of the model. The main steps are briefly described below.

3.1. Group Modeling and Data Collection

This step involved extensive gathering and analysis of statistical data (from 1990 to 2008 and projections) and information from multiple sources, including; Maui County Data Book (MCDB), Hawaii State Data Book (HSDB), Native Hawaiian Data Books (OHA), Energy Information Administration (EIA), District Water Services (DWS), Center for Water Resource Management (CWRR), Maui County Water Use and Development Plan, Department of Business, Economic Development and Tourism (DBEDT), State Department of Health (DoH), Maui County General Plan 2030, Maui Island Plan, and U.S Census reports.

Along with the detailed statistics collected from these sources, a significant portion of data pertaining to the water sector and technical information on the workings of specific systems were collected from a variety of USGS studies focusing on Maui and various aquifer systems within the region [9]. Many of these studies were very detailed reports outlining the peculiar interactions of selected water systems on Maui. Extracting and aggregating the relevant outcomes of these studies, and implementing them into a coherent and integrated model highlights one of the contributions brought by this study. This enables organizations such as MEDB to present scientific conclusions to the community in an understandable way to facilitate community conversations and to assist policy formulation and evaluation.

To promote the community aspect of this process, a group-modeling session was held on Maui that was attended by local stakeholders representing farmers, landowners, and other concerned citizens. Through this process, local needs and concerns were explored and included in the subsystem diagram of the water model that was jointly developed during group sessions, to then be incorporated into the actual model at a later stage. Group modeling session included the creation of a Causal Loop Diagram (CLD) that would represent the water sector (both demand and supply) of Maui County. Further the main drivers of demand and supply, including possible policies and actions were discussed and evaluated with local experts to better evaluate the role and relevance of both local cultural and structural constraints.

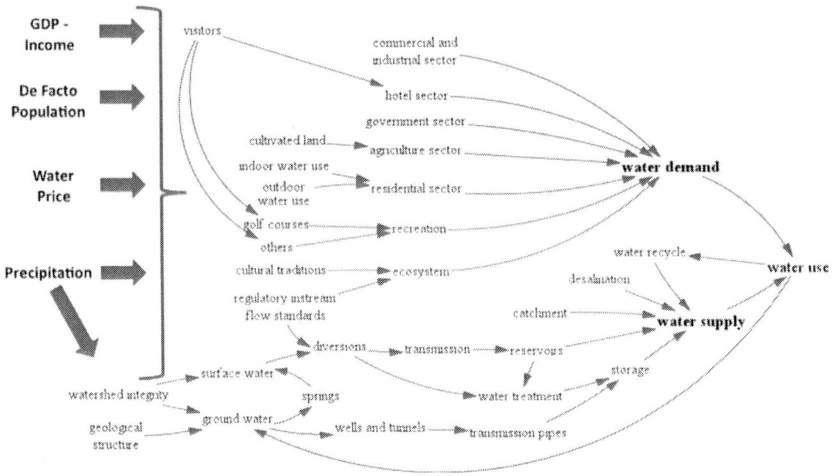

Figure 1. Causal Loop Diagram developed with the working group during group modeling sessions, main drivers of water demand and supply on Maui.

3.2. Model Development

The project employed a customized version of the PCM model developed by the Millennium Institute. System Dynamics was used as the foundation for the customization of the model, to enable the examination of complex, dynamic interrelationships between water and other social, economic and environmental sectors. Specifically, a detailed model of the water sector was constructed, to represent water demand and supply in Maui.

Scenarios relevant to the peculiarities of Maui are determined and processed to develop an understanding of the impacts of different policies and regulations being discussed within the community. While accounting for a decreasing trend in precipitation, as well as a reduction in the annual number of visitors, the model analyzes the impacts of instream flow regulations, increases in water efficiency and investing in increasing the supply of water (e.g., desalination).

4. MODEL DESCRIPTION

The PCM is constructed on a software platform that allows for the representation and interaction of a complex network of feedback loops, which underlie the mechanisms common to developing and industrialized countries [14]. Its structure is base don 20 years of research at the Millennium Institute and over 40 country applications of integrated national development models [20], such as PCM. The Maui model is organized in spheres, sectors and modules and was customized following a modular approach. PCM-Maui is largely focused on water and a more detailed description of non-water sectors is available in other publications (see for instance Bassi, 2009 [24]). Table 1 shows all of its spheres and sectors (with sectors being the building blocks of the higher level aggregation represented by the spheres).

Table 1. Modules, Sectors and Spheres of PCM-Maui Model.

Social Sphere	Population
Economic Sphere	Production
	Households Accounts
	Government Accounts
	Banks
Environmental Sphere	Land allocation
	Water demand and supply
	Energy demand and supply
	Air emissions

The content of individual modules is based upon accepted peer reviewed research within its designated field and targeted to specific governing dynamics isolated by the research inquiry. Also, all the inputs received during the working group meetings and group modeling sessions were directly incorporated in the model. The distinctive characteristic of PCM is the manner in which various determinants are linked together forming a complex link of feedback loops in which *ad hoc* research and/or policies can then be analyzed and weighted as driving or limiting the county's development agenda.

The core of the Economy sphere of PCM-Maui is represented by GDP, as one aggregated item, which is characterized by a Cobb-Douglas production function with inputs of labor and capital, visitor expenditure and total factor productivity, which includes energy prices. A Social Accounting Matrix (SAM) [25] is used to elaborate the economic flows between the main actors in the economy: government, households, producers and banks. The government sector collects taxes based on economic activity and allocates expenditures by major category, which then impacts the delivery of public services, subject to budget balance constraints. County budget categories are employed, and key macro balances are incorporated into the model.

The Social sphere contains detailed population dynamics by sex and one-year age cohort. These sectors take into account, for example, the interactions of family planning, health care and income on fertility and life expectancy, which in turn determines population growth. Population determines the labor force, which shapes employment. Employment and labor productivity affect the levels of production from a given capital stock.

The Environment sphere tracks land allocation, water and energy demand and supply, in addition to pollution (*i.e.*, emissions and water pollution due to desalination). It estimates the consumption of natural resources—both renewable and non-renewable—and can estimate the impact of the depletion of these resources on production and prices. Particular attention is devoted to water supply, where the unique peculiarities of Maui's geographical area and hydrological systems are incorporated into the analysis.

More specifically, the supply of water on Maui is largely influenced by rainfall. Precipitation impacts watershed integrity, which is responsible for the levels of surface and ground water in the county. Further, groundwater (GW) accumulates the flow of rain infiltration (GW_R)—which in turn is determined by the geological structure of the land, the flow of groundwater production (GW_P) and natural loss (GW_L):

$$D(GW)/dt = GW_R(t) - GW_P(t) - GW_L(t) \quad (1)$$

More specifically, GW_R is calculated using the total amount of water internally produced—which accounts for precipitation, multiplied by the fraction of rain evaporating immediately, and fog-, multiplied by the ratio of ground water recharge relative to precipitation. This methodology follows the analysis and data published by USGS [9]. GW_P instead is calculated by multiplying GW_R by the observed natural outflow share of water that infiltrates in groundwater reservoirs. This share is constant and exogenous. Finally, GW_L is calculated as the minimum between the maximum sustainable groundwater production and the difference between water demand and supply from surface water and desalination (in other words, groundwater production is the residual factor to satisfy demand, with the limitation of avoiding overproduction levels that would compromise the integrity of the reservoirs—a key concern for the local population).

Surface water (*SW*), which represents the stock of water above the ground, accumulates the flows of ground water emergence and recharge from rain (SW_I)—accounting for evaporation—as well as production (SW_P), diversions for agricultural irrigation (SW_{AI}), and the discharge in the ocean (SW_R):

$$d(SW)/dt = SW_I(t) - SW_P(t) - SW_{AI}(t) - SW_R(t) \qquad (2)$$

Surface and ground water flow through different channels, such as diversions, wells and tunnels, reservoirs and other forms of storage, which in turn define the actual available supply of water. Water from catchment systems, desalination and recycling plants also add to the total supply of water on Maui.

The demand for water (*WD*) comes from a variety of sectors, including, among others, the commercial and industrial (WD_P), residential (WD_R), religious institutions (WD_{RI}) and agricultural ones (WD_{AG}). It was discussed by some stakeholders that the amount of water that would preserve the local ecosystem be considered as demand, given its relevance. Ecosystem demand includes, but it is not limited to, the government regulated in-stream flow standards and the amount of water needed to preserve cultural traditions of the native farmers (e.g., taro and shrimp farming). Each sector has specific drivers that define the demand for water. For instance, residential water demand is mostly driven by population (*POP*) and per capita water use (WD_{PC}), precipitation (*P*) and water efficiency use (*WE*); agriculture water demand is influenced by the amount of cultivated, irrigated land, as well as rainfall, that also influences overall outdoor water use:

$$WD = WD_{AG}(t) - WD_P(t) + WD_R(t) + WD_{RI}(t) \qquad (3)$$

$$WD_R = ((POP(t) * WD_{PC}(t)) / (P(t)/ P(t_0))^{\delta(P)}) / (WE(t)/ WE(t_0)) \qquad (4)$$

The main drivers of demand within the system analyzed are GDP (or disposable income), combined with *de facto* population and precipitation.

5. SIMULATION RESULTS

The PCM-Maui model projects the social, economic and environmental development of Maui County starting from 1990 until 2030. The structure of the model, based on observed causal relations and correlations, and accounting for feedback loops, delays and nonlinearities, is compared to historical data from 1990 until 2008, to then project a baseline and an alternative scenario until 2030.

5.1. Baseline Scenario-Business as Usual

The baseline scenario, business as usual (BAU) assumes no changes in any of the current policies and regulations relating to water, and it mostly focuses on the future behavior of two key drivers of water demand and supply in Maui County, precipitation and visitors.

More specifically, the former is a proxy for the impact of global climate change on Maui and the effect it has on freshwater resources. Although the literature suggests that current precipitation models are inadequate at providing future precipitation levels [11], a 2004 USGS study has shown that from 1913 to 2002, there was a downward trend in annual rainfall over much of the area of Maui [26]. Since water resources are exclusively dependent on rainfall on Maui, any changes in the rainfall patterns and subsequent frequency and duration of droughts can affect both ground water and surface water supplies [11], through an increase in stream flows and groundwater recharge. Further, there is a significant correlation between precipitation levels and water demand on Maui. Because long-term trends in rainfall characteristics are difficult to identify, an exponential decline is assumed, and predictions of precipitation levels from 2010 to 2030 can be derived. At 2009 total precipitation in the County is at 2.49 trillion gallons/yr, approximately 3,103 mm/yr, decreasing towards 1.31 trillion gallons/yr by 2030, or 1,614 mm/yr.

The second assumption used to simulate the BAU case is the number of visitors reaching Maui County every year. As mentioned earlier, the economy of Maui is highly dependent on visitor numbers and tourism within the region. Accounting for almost 40% of economic activity on the island [2], changes in visitor levels have a strong impact on gross domestic product and the demand for water and other resources. Due to the economic crisis of 2008, the number of visitors to the region has declined considerably—by 18.45% over the first 5 months of 2009 [27]. Other sources, such as the MCDB, project future levels of annual visitors based on data from 2006, but fail to incorporate the economic crisis, resulting in overoptimistic predictions [1]. To account for the impact of the economic and financial crisis on Maui, the actual drop in annual visitors from 2007 to 2008, a loss of 374,300 annual visitors is inputted into the model, and the same growth rate projected by MCBD, 33% by 2030, is assumed.

Concerning the results of the simulation of the baseline scenario, social, economic and environmental indicators were selected to provide an overview of the projected development of Maui County. Table 2highlights the most important results of the baseline and alternative scenarios for selected years.

Table 2. Summary of findings of baseline and alternative scenario [1].

Variable	Unit	1990	2000	2010	2020	2030
De Facto Population	*people*	134,564	169,405	194,263	226,074	259,976
Agriculture Land	*acre*	406,248	404,049	403,133	401,811	400,518
Settlement Land	*acre*	25,000	27,388	28,315	29,752	30,976
Surface Water Production	*billion gallons/year*	4.54	3.84	2.76	2.18	1.61
GDP	*billion USD'00*	3.64	4.49	5.17	6.24	7.02
		3.64	4.49	5.17	6.25	7.03
PC disposable Income	*USD'00/people*	24,828	25,612	25,708	25,314	24,944
		24,828	25,612	25,708	25,329	24,958
Total Employment	*people*	58,440	69,856	81,234	89,394	93,031
		58,440	69,856	81,234	89,502	93,117
Annual Visitors	*million people/year*	2.39	2.34	2.30	2.71	3.12
		2.39	2.34	2.30	2.71	3.12
Water Demand	*billion gallons/year*	9.51	11.59	13.70	16.26	19.42
		9.51	11.59	13.57	14.73	16.20
Groundwater Production	*billion gallons/year*	4.96	7.75	10.94	14.07	15.25
		4.96	7.75	10.81	6.33	8.73
Production from Desalination	*billion gallons/year*	0	0	0	0	0
		0	0	0	6.21	5.86
Relative Groundwater Stock	*% relative to 1990*	100	96	88	78	66
		100	96	88	85	80
Water Runoff	*billion gallons/year*	344	295	232	199	166
		344	295	251	254	221
Energy Demand	*billion BTU/year*	18,888	29,763	40,363	49,467	63,850
		18,888	29,763	40,363	49,505	63,887
Agriculture Revenue	*million USD'00/year*	213	114	127	98	60
		213	114	124	88	59
Agriculture Employment	*people*	2,604	2,084	1,628	1,255	767
		2,604	2,084	1,587	1,132	760

[1] Clear region indicates baseline scenario, shaded region indicates alternative scenario.

Concerning the social sector, total resident population is projected to increase to 195,811 people in 2030, 36% above its 2008 level. Its age distribution shows a significant reduction of population between the age 15 to 19 and the age 20 to 24 cohorts—with the diminution amplified over time. There then seems to be an inflow at later ages, indicating the possible return of these students, potentially to pursue a career on Maui, or the inflow of population from the US mainland and other countries.

With economic activity in 2007 totaling $5.53 billion (in current USD), the decline in visitors is projected to cause GDP to fall to $5.16 billion USD in 2008, a 6.7% reduction. GDP is projected to recover by 2010, after a further decline in 2009, and then follow an average growth rate of approximately 1.7% per year.

Household revenue follows a similar trend, with a negative shock occurring with the 2008 economic crisis, and then steady growth afterwards. With a drop of $391 million USD, household revenue rises to $6.95 billion USD in 2010 (nominal), increases to $9.75 billion USD in 2020, and reaches $13.33 billion USD at 2030.

Conversely, per capita real disposable income tends to stay about constant over time. With a sudden drop in 2008, per capita real disposable income drops $2,065 (chained to 2000 USD) from 2007 to $24,800 '00USD in 2008. Disposable income then slightly increases over time to $24,944 '00USD in 2030.

Total employment appears to be following a decreasing exponential trend, as persons employed increases from 81,200 in 2010, to 89,300 in 2020 and leveling off to 93,000 people by 2030. The decreasing trend occurs due to the limiting factors within the economy of Maui, such as land available for agriculture, lower growth within tourism and capital available for investment, as well as for a higher reliance on technology and the services sector.

The amount of agriculture land is projected to slightly decrease over the simulation period. Historically, 406,248 acres of land was available for agriculture, whereas 404,049 acres were allocated to agriculture in 2000. Future projections indicate that cultivated land will be 400,518 by 2030. Differently, cultivated land, a subset of agriculture land, halved between 1990 and 2008, probably due to water stress (decreasing precipitation), declining profitability and higher competition—from abroad—in the agriculture business.

As discussed earlier, the reduction in precipitation due to climate change has a direct impact on water demand and supply. As agricultural, commercial and residential sectors develop and continue to use water, water demand is projected to increase as water available from precipitation declines. With a slight reduction in demand due to the recession of economic activity in 2008, demand grows to 13.7 billion gallons/yr in 2010, 16.3 billion gallons/yr in 2020 and 19.4 billion gallons/yr in 2030.

With reduced precipitation feeding groundwater and surface water levels, the total supply of water on Maui is reduced over time, as depicted by the projection of the amount of water in aquifers, relative to its 1990 level (see Table 2).

With the projected growing demand for water, supplies of water are projected to eventually reach a maximum level and begin to decline as stocks are eventually depleted in the 2 main aquifers currently used for groundwater production. This is consistent with a variety of studies that project demand to surpass supply by 2020 [2,3,11,12]. Although, due to the fact that these studies do not incorporate recent changes in economic activity, such as the reduction in visitors caused by the 2008 economic crisis, the point at which demand exceeds supply at 2020 seems prudent. By incorporating the reduction in visitors and the consequent reduction in gross domestic product, our model projects that the supply of water will reach a maximum at around 2024 and begin to level off as the main aquifers begin to be chronically depleted below their sustainable level.

The baseline scenario assumes that no investment will be allocated to the exploitation of aquifers in the Hana region or in other areas where the capital cost of such a project would be overwhelming and restrictive. With 13.7 billion gallons/yr in 2010, the total supply of water reaches a maximum at 17.2 billion gallons/yr in 2024, and then decreases to 16.8 billion gallons/yr in 2030. The maximum limit at which these aquifers can be exploited has been assumed to be the sustainable yield indicated by the Center for Water Resource Management, which is about 40% above current levels. It is assumed that once reaching this limit, aquifers will begin to degrade, the system in which they are contained will break down and the level at which they can be pumped will diminish [26].

5.2. Alternative Scenario—Implementing Policies and Regulations

After having investigated the outcome of the baseline scenario, a variety of policies and regulations are incorporated to analyze the change and impact on future projections. There are three policies that are simultaneously implemented within this scenario.

The first is the instream flow regulations that have been widely discussed within the County of Maui, and have become a contentious issue. The instream flow standards recommended by the Commission on Water Resources Management include the streams such as Honopou being capped at 1.79 Mgal/day, Hanehoi at 1.72 Mgal/day, Piinaau at 3.56 Mgal/day, Waiokamilo at 3.17 Mgal/day and Wailuanui at 1.97 Mgal/day. These are just 5 streams that have been selected within a 27-

stream system, which is part of a network totaling over 100 streams. For the purposes of the simulation, the average value over the 5 streams was calculated and knowing that more streams will be regulated over time, it is assumed that the amount of streams being imposed with these standards will be 60 streams.

The second policy implemented into the analysis is the increase of water use efficiency (or voluntary conservation) over time. As time progresses, new practices emerge and technology processes improve to allow for the more efficient distribution and consumption of water, as well as an increase in conservation of the resource. With diversion infrastructure having been implemented over 100 years ago, there is a considerable amount of loss within the system, as well as inefficiencies throughout the process. Moreover, it is reasonable to assume that technologies or practices will be implemented that reduces the necessity for water at the residential, commercial and agricultural levels over time. Within this simulation, a 20% increase in water efficiency by 2030 is incorporated into the model.

The last policy that is measured within the model is the scenario of investing in and developing a desalination plant on Maui. An option that is being considered on Maui, a desalination plant would allow for the increase of the water supply by treating brackish water through reverse osmosis. For the purposes of the model, it is assumed that a plant that can bring the levels of the Central Aquifer System to its sustainable yield. As mentioned earlier, the Aquifer is currently being pumped at 168% above its sustainable yield, which translates into 18 Mgal/day above its threshold. Currently, a desalination plant on Maui with an expandable capacity of 5 Mgal/day of useful water is being discussed, with capital costs ranging around $50 million USD [28,29]. Assuming that the excess 18 Mgal/day is satisfied with these plants, a $180 million USD investment is required in 2010 to increase water supply by 18 Mgal/day by 2012. It is assumed that it is a one-time investment, requiring two years to reach full capacity and having a capital lifetime of 30 years.

When running the simulation, the first variable to analyze is the effect these policies have on the demand of water. Within this scenario, the demand for water decreases from the baseline case. Due to the increase in efficiency, less water is demanded for the same rate of use. With a 20% increase in efficiency by 2030, there is a 5.3% drop in demand in 2015, a 9.4% reduction in 2020 and a 16.5% decrease—16.2 billion gallons/yr vs. 19.4 billion gallons/yr, by 2030.

Water supply in the alternative scenario will be mostly attained from desalination and groundwater use. Groundwater production, as a share of total supply, moves from 80% in 2010 in the BAU, to 35% in 2012 thanks to the investment in the desalination plant. Due to increasing water demand over time, groundwater production grows back to 54% by 2030. Desalination water production is able to satisfy about 45% of water demand in 2012 (74% of residential use), with the remaining 20% being produced from surface water. The figures above refer to the total supply of water satisfying demand from the residential, commercial, industrial, religious and agriculture sector, excluding the amount needed for irrigation, following the classification of the USGS study on groundwater availability, as well as the 2007 Water Use and Development Plan on Maui.

An interesting outcome of this scenario on water supply is the reduction of groundwater production below the maximum sustainable level of the aquifers currently being overproduced. As indicated within the baseline scenario, the supply of water is capped at the sustainable yield, and supply begins to diminish after 2024, once the maximum level is reached. With this investment to increase the supply of water, that maximum is not reached within the simulation period, and remains within the sustainable yield throughout the simulation horizon.

As more of the fraction of groundwater production is satisfied by desalination, there is a cost associated with implementing such a decision. Firstly, there is an increase in electricity demand, caused by the estimated 0.015 kWh/gallon used by the desalination plant. Total electricity demand increases initially by 6.26 GWh/yr in 2011 as the desalination plant comes online, and peaks at 7.16% above the baseline scenario at 199.81 GWh/yr of energy demanded in 2012. Since the capital efficiency of the plant decreases over the 30 years, a slight reduction in energy consumption is experienced over time and there is only a 5.02% increase in energy demand compared to the baseline scenario in 2030. This translates into an increase of 20 MW of power generating capacity, which can cost between $10–40 million, depending the energy source considered.

Furthermore, there are additional costs associated with building a desalination plant to satisfy growing demand on Maui. When water is processed, liquid waste is created that contains toxic components. With high salt concentrations and other chemicals used within the process, it is estimated that for every gallon of useable water generated, 1.4 gallons of polluted water are created. This results in 8.7 billion gallons/yr of wastewater generated at the peak of production in 2013, and approximately 167 billion gallons of wastewater created throughout 2030.

Lastly, the initial cost of the desalination plant is a major factor to take into account. Currently, $180 million USD is approximately 40% of planned government expenditure (FY 2009–2010) and 15% of private investment in 2008, which shows the scale of undertaking such a project. If actions are taken to increase water supply to satisfy exceeding demand, funding for such developments must be readily available, which may not be the case.

The regulation of instream flow has become a contentious issue on Maui due to the diversion of natural stream systems for agriculture in the late 1800s, as mentioned earlier. By increasing the level of water within the streams, less water is being diverted for agriculture, thus decreasing the amount available for irrigation. By imposing this regulation, there will be a reduction of 2.5% of water allocated to irrigation compared to the baseline scenario in 2010. This will increase to 9.8% in 2020, and reach a 13.98% decline by 2030. Thus, 54.74 billion gallons/yr are diverted by 2030.

The impact on the agriculture sector is indicated by the change in revenue compared to the baseline scenario. In 2010, revenue is projected to decrease by 2.51%, or $3.2 million USD'00, due to lower water availability, and reaches a reduction of 9.87%, or $9.62 million USD'00 in 2020. By 2030, the reduction in revenue is only 0.93%, at $560,272 USD'00. It should be noted that due to the decreasing trend in precipitation over time, the agriculture sector if projected to feel pressure throughout the simulation period, and the implementation of instream flow regulations only marginally decrease projected revenues. The decreasing rate of revenue reduction within the trend can be associated with the increase in water efficiency use—which limits the impact of decreasing water availability, as the agriculture sector reduces water demand over time as technology and maintenance of the diversion system allows for the more efficient use of water.

When analyzing employment levels in the agriculture sector within the alternative scenario, jobs are strongly correlated and influenced by the availability of water. As water levels decrease, the amount of jobs lost in the sector increase. In 2010, 41 jobs are lost, a reduction of 2.5%. In 2020, 123 jobs are lost, at which point the decline reaches its maximum rate in 2023 and then begins to subside. This can be attributed to the increase in energy efficiency causing a reduction in water demand over time, which results in a smaller fraction of the baseline scenario lost; only 7 jobs lost in 2030, or a 0.98% reduction.

Of difficult analysis are the potential impacts of water availability on specific locations and companies, such as HC&S. In fact, it can be reasonably expected that the company would be forced to close operations before its water supply reduces to zero. In this respect, the identification of a tipping point would require more careful research but would certainly add value to the analysis, as HC&S employs about 800 people on Maui and largely contributes to the GDP generated by the agriculture sector.

6. CONCLUSIONS

The purpose of this study was to analyze the impact of water policies and interventions within the County of Maui. Due to a strong cultural relevance of the stream system and water availability within Maui, there have been extensive community discussions on regulating instream flows, as well as implementing new technologies that would help to satisfy increasing demand.

Being essential to the local economy, water is an integral part of all systems and processes—policies and regulations affecting the future availability of this resource have system-wide ramifications. To understand these impacts, an integrated approach was employed, utilizing System Dynamics to create a model that is highly customized to Maui County.

Furthermore, because water has very significant implications on the community at large, the management and regulation of this resource has become a contentious issue. Interest groups representing local industry, environmental groups and the general community have been promoting their respective ideals, while remaining uninformed of the larger—possibly unintended—consequences.

Thus, the goal of the model is to provide a foundation on which local stakeholders and the general community can both achieve a consensus on current issues, develop an understanding of the systemic consequences of their actions, and to test and evaluate proposed polices and regulations. By empowering both policymakers and the layperson with facts about the state of the situation, as well as identifying key cross-sectoral links—conversations and debates move from interest groups confronting one another (often discussing about the validity of one assumption versus another), to community dialogues that develop consensus on how to solve the problems and achieve a sustainable solution. This goal was achieved by (1) involving local experts and other interested parties in the modeling process, drawing the underlying structure of the water sectors of the model in CLDs; (2) incorporating this in the stock and flow diagram (PCM-Maui) and cross checking data availability and its consistency with all interested parties; (3) organizing ...king sessions to review the model and evaluate its results, simulating ... of scenarios to accommodate requests from the working group.

This paper analyzes the impact of three policies and actions in light of the baseline scenario, while also incorporating a decline in precipitation and projected annual visitors during the simulation period. These policies include: setting instream flow standards to meet current regulations, assuming efficiency levels for water use increase over time, and investing in and developing a desalination plant to satisfy excess demand.

Policies such as increasing water use efficiency and desalination represent actions that the community can take to increase the supply of water over time. These have direct implications on the supply and demand of water, but also contain indirect costs if utilized. Firstly, the demand for water over time will decrease due to the more efficient use of the resource. Moreover, the supply of water will be able to increase without reaching its sustainable yield, due to the reduced stress on groundwater aquifers because of water being produced from the proposed desalination plant.

There are indirect costs associated with these actions though. With the development of a proposed desalination plant on Maui that facilitates the excess demand currently experienced, there will be a rise in energy demand and an increase in the necessary power generation capacity in the long term. With the County of Maui experiencing some of the highest energy costs per capita in the nation, the increase in energy demand will need to be satisfied and could prove costly in the long term. Furthermore, there is a considerable amount of wastewater created by a desalination plant, and this could have negative ramifications on the delicate structure of the local ecosystem.

The instream flow standards being discussed also have system wide ramifications, with indirect effects on the agriculture industry. Taking into account the decline in the industry due to the reduction in precipitation over the simulation period, there is a further negative impact on revenue and employment within the sector. Even though agriculture comprises 2% of the gross domestic product, the industry has had its roots on Maui for over a century, thus there are strong social implications of negative ramifications on the sector.

Due to this outcome, and other consequences, which have implications on cultural sustainability, there needs to be a community dialogue to discuss these policies and arrive at common solutions to help facilitate the transition towards a more sustainable future. PCM-Maui can provide valuable insights to policy makers and local stakeholders, but can also be used as an effective tool to create and facilitate a channel of communication within the community. Furthermore, the MI-MEDB approach is unique in that the *ad hoc* modeling process will be acquired and maintained by MEDB upon completion. By handing over ownership to the local partner, the community retains possession of the model, and with skills developed in training sessions, MEDB is able to update and customize the model further. This allows the model to become a tool that is not only developed by the community, but also owned and maintained by it as well.

By combining this dialogue with the holistic framework based on the initial framework of PCM, AQUA and well targeted customizations, the PCM-Maui is a highly customized tool to inform policy formulation and evaluation in Maui County. Not only will the model allow for the analysis of a variety of policy options, providing insights on the impacts of various water allocation strategies within the region, but the approach will also allow for the analysis of the cross-sectoral impacts and synergies within Maui County.

REFERENCES

1. *Maui County Data Book (MCDB) 1995–2008. County of Maui Office of Economic Development. Hawai'i Small Business Development Center Network*; Hawai'i Business Research Library: Maui, HI, USA, 1995–2008.
2. Maui County General Plan Advisory Committee. *Maui Island Plan (MIP), 2008,* Available online: http://www.co.maui.hi.us/index.aspx?NID=1120 (Access ed on 4 August 2009).
3. Socio-Economic Forecast (SEF). *The Economic Projections for the Maui County General Plan 2030*; Maui County Planning Department: Maui, HI, USA, 2006.
4. Lau, S.L.; Mink, J.F. *Hydrology of the Hawaiian Islands*; University of Hawaii Press: Maui, HI, USA, 2006.
5. Stearns, H.T.; Macdonald, G.A. *Geology and Ground-Water Resources Of The Island Of Maui*; Hawai'i Division of Hydrography Bulletin: Maui, HI, USA, 1942.
6. Giambelluca, T.W.; Nullet, M.A.; Shroeder, T.A. *Rainfall Atlas of Hawai'I*; Technical Report R76; Hawai'i Department of Land and Natural Resources, Division of Water and Land Development: Maui, HI, USA, 1986.
7. Warzecha, R.L. East Maui Irrigation Company, Past Present and Future. In Proceedings of a Conference Exploring Maui Water

and Energy: Future Needs and Directions, Maui, HI, USA, December 14; 1986.

8. Cataluna, D.B. Water Development and Use at Wailuku Sugar. In Proceedings of a Conference Exploring Maui Water and Energy: Future Needs and Directions, Maui, HI, USA, December 14; 1986.

9. Engott, J.A.; Vana, T.T. *Effects of Agricultural Land-Use Changes and Rainfall on Ground-Water Recharge in Central and West Maui, Hawaii, 1926–2004*; Technical Report 2007-5103. U.S. Geological Survery Scientific Investigations: Island of Maui, HI, USA, 2007. Available online: http://pubs.usgs.gov/sir/2007/5103/ (accessed on August 4 2009).

10. Meyer, W.; Todd, K.P. *The Response of the Iao Aquifer to Groundwater Development, Rainfall, and Land-Use Practices between 1940 and 1998*; Technical Report 00-4223; U.S. Geological Survey Water-Resources Investigations: Island of Maui, HI, USA, 2001.

11. *Water Resource Protection Plan (WRPP), Hawaii Water Plan, 2008*; Commission on Water Resource Management, Department of Land and Natural Resources, State of Hawaii: Maui, HI, USA, 2008.

12. Freedman, C. *Maui County Water Use and Development (WUDP)*; Water Use and Demand Department of Water Supply Systems: Maui, HI, USA, 2007.

13. Witney, L.D.; Bowers, F.A.I.; Takahashi, M. *Taro Varieties in Hawaii; Bulletin No. 84*; College of Tropical Agriculture and Human Resources, University of Hawaii: Manoa, HI, USA, 2007.

14. Pedercini, M.; Kopainsky, B.; Davidsen, P.L.; Alessi, S.M. Blending Planning and Learning for National Development. In Proceedings of the 2008 International Conference of the System Dynamics Society, Boston, MA, USA, July 29–August 2; 2007.

15. Stern, N.H. *Great Britain Treasury. The Economics of Climate Change: The Stern Review*; Cambridge University Press: Cambridge, UK, 2007.

16. Gilmour, J.K.; Letcher, R.A.; Jakeman, A.J. Analysis of an integrated model for assessing land and water policy options. *Math. Comput. Simulat.* **2005**, *69*, 55–77.

17. Sterman, J.D. A skeptic's guide to computer models. In *Managing a Nation: The Microcomputer Software Catalog*; Barney, G.O.,

Kreutzer, W.B., Garrett, M.J., Eds.; Westview Press: Boulder, CO, USA, 1988; pp. 209–229.

18. Dinar, A.; Letey, J. *Modelling Economic Management and Policy Issues of Water in Irrigated Agriculture*; Praeger Publishers: Westport, CT, USA, 1996.

19. Griffin, R.; Hsu, S. The potential for water market efficiency when instream flows have value. *Am. J. Agric. Econ. Assoc.* **1993**, *75*, 292–0303.

20. Bassi, A.M.; Baer, A.E. Quantifying cross-sectoral impacts of investments in climate change mitigation in Ecuador. *Energy Sustain. Dev.* **2009**.

21. Hoekstra, A.Y.; Beusen, A.H.W.; Hilderink, H.B.M.; van Asselt, B.A. *Perspectives on Global Change: A TARGET Approach*; Cambridge University Press: Cambridge, UK, 1997; Chapter 14.

22. Winz, I.; Brierley, G. The Use of System Dynamics Simulation in Integrated Water Resources Management. In Proceedings of the 25th System Dynamics Conference, Boston, MA, USA, July 29–August 2; 2007.

23. *Focus Maui Nui Final Report (FMN)*; Maui Economic Development Board: Maui, HI, USA, 2003.

24. Bassi, A.M. *An Integrated Approach to Support Energy Policy Formulation and Evaluation*; Univeristy of Bergen: Bergen, Norway, 2009.

25. Drud, A.; Grais, W.; Pyatt, G. Macroeconomic modeling based on social accounting principles. *J. Policy Model.* **1986**, *8*, 111–145.

26. Oki, D.S. *Trends in Streamflow Characteristics at Long-Term Gauging Stations, Hawaii*; Technical Report 2004-5080; U.S. Geological Survey Scientific Investigations: Island of Maui, HI, USA, 2004.

27. *DBEDT Data Warehouse (Economic Development and Tourism, Input/Output Model)*, Research and Economic Analysis Division, Department of Business, Economic Development & Tourism, Government of Hawaii: Honolulu, HI, USA, 2008; Available online: http://www.hawaiieconomicdata.com/ (Accessed 10 August 2009).

28. Leone, D. Kapolei desalination project gets under way. *Star Bulletin*, August 29 2003. Available online: http://archives.starbulletin.com/2003/08/29/news/index4.html (accessed on 27 July 2009).

29. Zhoua, Y.; Tolb, R.S.J. Evaluating the Costs of Desalination and Water Transport. Available online:http://www.fnu.zmaw.de/fileadmin/fnu-files/publication/working-papers/Desalination FNU41_revised.pdf (accessed on 27 July 2009).

9
CHAPTER

Emerging and Innovative Techniques for Arsenic Removal Applied to a Small Water Supply System

António A.L.S. Duarte [1,*], Sílvia J.A. Cardoso [2] and António J. Alçada [2]

[1]Department of Civil Engineering, School of Engineering, University of Minho, Largo do Paço, 4704-553 Braga, Portugal

[2]Águas do Zêzere e Côa, SA, Rua Dr. Francisco Pissarra de Matos, 21-r/c, 6300-906 Guarda, Portugal

ABSTRACT

The impact of arsenic on human health has led its drinking water MCL to be drastically reduced from 50 to 10 ppb. Consequently, arsenic levels in many water supply sources have become critical. This has resulted in technical and operational impacts on many drinking water treatment plants that have required onerous upgrading to meet the new standard. This becomes a very sensitive issue in the context of water scarcity and climate change, given the expected increasing demand on groundwater sources. This work presents a case study that describes the development of low-cost techniques for efficient arsenic control in drinking water. The results obtained at the Manteigas WTP (Portugal) demonstrate the successful implementation of an effective and flexible process of reactive filtration using iron oxide. At real-scale, very high removal efficiencies of over 95% were obtained.

Keywords: safe drinking water; public health; arsenic removal; emerging techniques; real-scale removal efficiencies; water sources sustainability; Manteigas WTP (Portugal)

1. INTRODUCTION

Arsenic in drinking water has been reported as the most widespread geogenic contaminant in water sources worldwide. Groundwater contamination is of global concern and arsenic-associated human health problems have now been recognised in many parts of the world, mainly in developing countries [1]. A wide variety of adverse health effects, including severe skin lesions, cardiovascular and haematological effects, and neurological disturbances effects have been attributed to chronic arsenic exposure, primarily from drinking water [2]. Furthermore, several epidemiological studies have confirmed that chronic arsenic poisoning causes skin and internal cancers [3].

Considering the lethal impact of arsenic on human health, environmental authorities have taken a more stringent attitude towards the presence of arsenic in water. In 1993, the World Health Organization (WHO) had recommended a maximum contaminant level (MCL) of arsenic in drinking water of 10 ppb [4]. The WHO recommendation was adopted by the EU in 2003 (Directive 98/83/EC), thereby revoking the previous 50 ppb limit. The new MCL was later transposed to the Portuguese legislation through Law Decree (DL) no. 236/2001. The drastic reduction of the arsenic MCL from 50 to 10 ppb has led many impoundments, which serve small and medium water supply systems, to become critical for this contaminant. Consequently, drinking water facilities are undergoing several technical and operational changes induced by the non-compliance (though seasonal) of raw water arsenic levels with the new quality standard. These changes concern:

- many Water Treatment Plants (WTP) that require upgrades to address arsenic removal in order to comply with the new lower limit;
- many drinking water supply systems managers that need to build new plants with arsenic removal facilities, since this contaminant has now become a critical parameter.

When arsenic contamination is identified and quantified to be above the MCL, managers are confronted with either finding other water sources or implementing arsenic removal operations. When a safer drinking water source is not available or it becomes too expensive to exploit—one that is both low-arsenic or arsenic-free, and exhibits acceptable microbiological quality-treating raw water for arsenic removal is often the sole viable option to explore. In this case, there is ample justification for the development of innovative removal technologies that are more efficient and economically sustainable for small and medium-sized water supply systems. This issue is very sensible in the context of water scarcity and climate change.

The work presented herein summarises the major processes (conventional and emerging) that can be used for arsenic removal in drinking water treatment, including an analysis of corresponding efficiencies, in order to establish selection criteria of those technologies as a function of the raw water characteristics and/or treatment schemes for existing WTP.

In this context, the authors present a case study describing the rehabilitation of the WTP of Manteigas carried out by the *Águas do Zêzere e Côa* (AdZC) Company, concerning the development and installation of a suitable arsenic removal facility [5]. Their process decisions and methodological design allowed the managing company to avoid the rash acquisition of an expensive and pre-formatted arsenic removal solution.

2. ARSENIC TOXICITY AND RELATED HEALTH HAZARDS

Arsenic naturally occurs in over 200 different mineral forms. Of these, approximately 60% are arsenates, 20% sulphides and sulphosalts; while the remaining 20% include arsenide, arsenite, oxides, silicates and elemental arsenic [6]. Arsenopyrite is produced by hydrothermal activity associated with the intrusion of granitic magma and orogenesis.

Arsenic pollution of natural waters has become an international sanitation problem that currently affects over 40 million people in the World. It was initially reported in Bangladesh and in some countries of Latin America, where groundwater arsenic concentrations surpass 3.4 mg/L (e.g., in Córdoba, Argentina). In New Zealand, Romania, the Russian Federation, Spain and the USA, arsenic levels between 0.4 and 1.4 mg/L have been reported for carbonated water springs. In Taiwan, artesian aquifers display concentrations above 1.8 mg/L. In Portugal, water sources that exhibit higher concentrations of arsenic (approximately 800 ppb for groundwater and 60 ppb for surface waters) are generally located in Trás-os-Montes and Alto Douro [7], where the presence of arsenic-rich quartz-sulphur minerals is very common. Minho, Beiras, Ribatejo and Alentejo are additional locations where the legal contaminant concentration (10 ppb) is now exceeded. This has justified the pertinence of epidemiological studies in exposed populations (over many years) to the ingestion of waters containing arsenic concentrations between 10 and 50 ppb (the new and the old legal limit). These studies would allow the evaluation of the real exposure impacts on public health, thereby providing a valuable contribution and support to future and possible proposals for revising and changing the arsenic quality standards for drinking water [8].

Anthropogenic sources of arsenic may result from processing of a variety of ores (e.g., copper, gold, nickel, lead and zinc), from wool and cotton processing, from ingredients of many insecticides and herbicides, from additives to various metal alloys, from mining, from hazardous waste sites, or from the glass and semi-conductor industry. In general, groundwaters exhibit higher concentrations of arsenic species that are more toxic than those found in surface waters. It has been found that when river water—a primary source of drinking water—is polluted by industrial or mining effluents, or by geothermal waste, the arsenic concentration increases [9].

Acute and sub acute poisoning results from ingestion of large quantities of arsenic with lower exposure time, whereas chronic poisoning occurs due to consumption of arsenic contaminated water for a long time period.

Most cases of acute arsenic poisoning occur from accidental ingestion of insecticides or pesticides, being urinary arsenic concentration the best indicator of recent ingestion (1–2 days). The lethal dose of arsenic in acute poisoning ranges from 100 mg to 300 mg. Nonspecific gastrointestinal effects such as diarrhoea and cramping, haematological abnormalities including anaemia and leukaemia, peripheral neuropathy (similar to Guillain-Barré syndrome), renal failure, respiratory failure and pulmonary oedema are common features of acute poisoning, which may lead to shock, coma, and even death. Depending on the quantity consumed, death usually occurs within 24 hours to four days [10]. Metabolic changes (such as acidosis, hypoglycaemia and hypocalcaemia) with acute arsenic poisoning are also reported.

Long term arsenic toxicity leads to multisystem disease and the most serious consequence is malignancy. The clinical features of arsenic toxicity vary between individuals, population groups, and geographic areas. It is unclear what factors determine the occurrence of a particular clinical manifestation or which body system is targeted. Thus in persons exposed to chronic arsenic poisoning, a wide range of clinical features are common. The onset is insidious with non-specific symptoms of abdominal pain, diarrhoea, and sore throat. The skin is quite sensitive to arsenic and dermatological changes are a common feature related to long term exposure. Skin lesions (hyperkeratosis and dyspigmentation) have been observed even in cases of exposure to levels in the range of 5–10 ppb [1]. Arsenic associated skin cancer, Bowen's disease, is an uncommon manifestation in Asians and may be due to the high skin melanin content and increased exposure to ultraviolet radiation. Arsenic may cause a basal cell carcinoma in a non-melanin pigmented skin. The latent period after exposure may be as long as 60 years. In Latin America, chronic endemic regional hydroarsenism—*Hidroarsenicismo Crónico Regional Endémico* (HACRE)—manifests itself by the onset of neurological disorders and severe skin lesions (Figure 1).

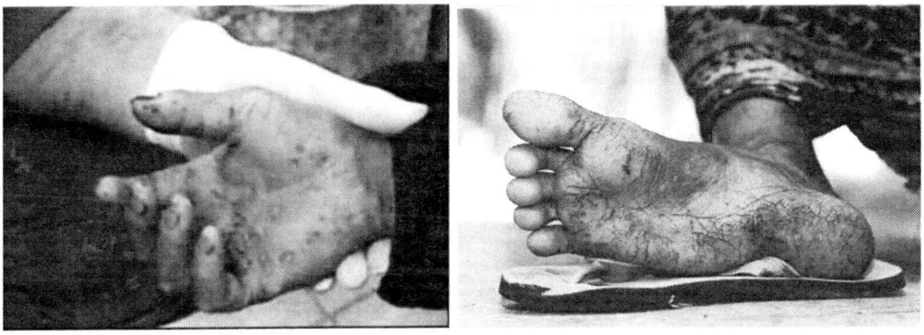

Figure 1. Skin lesions caused by prolonged ingestion of arsenic.

The conclusions of several epidemiological studies confirmed the potentially carcinogenic effect (skin, lung, bladder, kidney, liver, uterus and gallbladder) of a few inorganic species of arsenic when present in high concentrations. This led the WHO to recommend in 1993 a more restrictive guidance value of 10 ppb as the quality standard for drinking water, a value that is five times lower that the previous recommended limit. This drastic reduction in the maximum arsenic limit issued by the WHO has an essentially preventive character, since that parametric value is not yet sufficiently supported by extensive and conclusive epidemiological studies that are urgently warranted. Later on, in 2003, the United States Environmental Protection Agency (USEPA) drafted a proposal issuing final guidelines for cancer risk assessment [11]. The revised document advocates the use of nonlinear relationship between arsenic carcinogenesis and its dose in drinking water.

Since the removal of arsenic from raw water is often the only viable option in order to obtain safe drinking water, it is pertinent to globally intensify applied research efforts. These should address both the quantification of arsenic effects in health (toxicity levels) and the development of innovative technologies for arsenic removal that can be more efficient and sustainable, especially for small water supply systems in rural areas.

3. TECHNOLOGIES FOR ARSENICAL REMOVAL

Arsenic removal technologies should meet several basic technical criteria, including robustness, not causing any undue adverse effect on the environment, the ability to provide supply systems with the capacity to produce water in adequate quantity across different seasons and taking into consideration present climate change scenarios, and with the required physical-chemical and microbiological quality.

The presence of high concentrations of arsenic in water sources ought to be analysed in two distinct stages. The first stage consists in verifying the existence of safe alternative sources with very low or null arsenic concentrations that also exhibit adequate physical-chemical and bacteriological quality. In case those exist, the hypothesis of dilution should be pondered by means of conducting an option-specific economic study. When safer, alternative locations are not available or are scarce, thereby preventing the switch of water intakes from arsenic-polluted sources, arsenic removal technologies—conventional and emerging—ought to be used, even if it implies a seasonal operation, e.g., only in periods when the regulation parametric limit is exceeded.

The second stage consists in the definition of the conceptual scheme to adopt for arsenic removal considering the eventual presence of arsenite in the water source. Where the removal is more complex, there are two distinct removal alternatives to consider: carry out a preliminary oxidation step (through conventional or alternative processes) prior to arsenate removal, or use a membrane filtration operation (reverse osmosis or nanofiltration), excluding the arsenite oxidation step [12]. The success of the application of an arsenic removal technique depends on the following factors: parametric value to achieve; raw water arsenic concentration; flow rate to treat; regional lithography; water source and existing treatment scheme [13]. The judicious selection of the most adequate alternatives should be based on a cost-benefit analysis conducted for each one of the possible removal techniques.

To deal efficiently with arsenic removal, regular monitoring is required to identify the factors that control arsenic concentration and arsenic speciation. Such monitoring programmes are not easily established in developing countries and thus, arsenite-oxidizing processes integrated into arsenic removal methods appear as a safety precaution measure in high-risk areas [14].

3.1. Conventional Technologies

All of the conventional technologies for arsenic removal rely in a restrict group of basic physical-chemical processes [15]. These include:

- Oxidation/reduction reactions, which reduce or oxidise chemicals, altering their chemical form but not removing arsenic from solution (often used to optimize other processes);
- Precipitation, which causes dissolved arsenic to form a low-solubility solid mineral that can be removed through sedimentation and filtration. When coagulants are added, other dissolved compounds can form solids (known as co-precipitation);
- Adsorption and ion exchange using various solid materials such as iron and aluminium hydroxide flocks, which have a strong affinity for dissolved arsenic. Ion exchange can be considered as a special form of adsorption;
- Physical exclusion, which uses synthetic membranes that are permeable to certain dissolved compounds but exclude others, and that can act as molecular filters to remove dissolved arsenic along with many other dissolved and particulate compounds.

Coagulation-filtration and softening with lime addition techniques are economical, but display lower efficiencies (<90%). Among the most efficient techniques (>95%), adsorption using aluminium carbonates stands out as one of the least expensive [16]. Several inorganic-organic hybrid adsorbents bearing *thiol*groups have been prepared by modifying activated alumina (AA) with mercaptopropyl-functionalized silica under different experimental conditions. Compared with AA, hybrid adsorbents enhance adsorption capacities for arsenite due to the introduction of *thiol* groups, and still maintain the merit of the AA for arsenate adsorption [17]. However, hybrid adsorbent must be selected by considering not only the adsorption capacity but also the environmental-friendliness and cost-effectiveness of production. Table 1 summarises the removal efficiencies that are normally reported and the main factors responsible for the reduction of those efficiencies.

Table 1. Efficiency of conventional arsenic removal techniques.

Technology	Chemical reagent	As (III) removal efficiency (%)	As (V) removal efficiency (%)	Ideal conditions
Coagulation-filtration precipitation (including lime softening)	Ferric chloride	<30	90–95	pH 6–8
	Sulphates (aluminium, copper, ammonia)	<30	80–90	pH 6–6.5
Adsorption	Activated carbon or activated alumina	30–60	>95	pH 5.5–6
	Iron hydroxide (granular)	30–60	>95	pH near 8
Ion exchange	Anionic resins	<30	80–95	$[SO_4^{2-}] < 20$ mg/L $[SDT] < 500$ mg/L
Membrane filtration (nanofiltration and reverse osmosis)	-----	60–90	>95	Presence of dissolved *As*
		80–95	>95	

Most of these techniques have been used in arsenic removal at WTPs as adequate solutions for small (500 inhabitants) or large (>1,000,000 inhabitants) water supply systems. However, they show some limitations: addition of chemical products to the water might require complementary treatment; possibility of ion interference (e.g., sulphates) in the removal efficiency; and high installation and operation costs associated with low efficiencies.

Despite its effectiveness, the use of membranes for arsenic removal significantly heightens (approximately 10 times more) the unit costs of the process, often rendering it an unsustainable solution for small water supply systems.

3.2. Emerging Physical-Chemical Technologies

The conventional technologies for arsenic removal referred above are fairly well documented, although some of the systems have only come into prominence in recent decades. Nowadays, extensive research has been conducted towards identifying new technologies for arsenic removal. These focus particularly on low-cost systems that can be applied to small water supply systems, in order to increase efficiency and improve the cost-benefit balance of arsenic removal.

Technology research has also focused on the improvement of existing conventional technologies such as adsorption [18,19], by modifying or using novel adsorbent materials [20,21], or by introducing new chemical oxidation processes. Most of these technologies rely on the oxidation of arsenite followed by filtration through a porous material, where arsenic is removed through adsorption and co-precipitation.

TiO_2 immobilization, on a PET surface, combined with co-precipitation of arsenic on iron (III) hydroxides (oxides), could be an efficient way for total inorganic arsenic removal from waters [22,23]. Because of their very strong affinity for arsenic, iron compounds are used by many removal systems. This is also the case for the application implemented and described in the case study presented below.

Recent research work was developed to find a suitable iron (II) to arsenic ratio in water to reduce arsenic to 5 ppb (or lower) through slow sand filtration. It was found that a ratio of 40:1 was necessary to ensure the desired arsenic concentration in the treated effluent [24].

3.3. Emerging Biological Technologies

Biological removal processes by indigenous bacteria can play an important role in catalyzing many of the above chemical processes, but relatively little is known about the potential for biological removal of arsenic from water.

Depending on the physical-chemical condition of the environment, some arsenic compounds can be highly soluble, resulting in a high level of bioavailability. Its toxicity and bioavailability depend on its speciation, which in turn, depends on microbial transformations, including reduction, oxidation and methylation. Biological and microbiological methods can represent economically viable as well as environment-harmless alternatives for arsenic removal. Active treatments of arsenic-contaminated waters benefit from the knowledge of arsenic bacterial metabolism. However, many studies were confined to laboratory and need to be tested under real-scale conditions to assess their viability.

The biogeochemical cycle involves several physical-chemical processes (such as oxido-reduction, precipitation/solubilisation, and adsorption/desorption processes) as well as biological mechanisms, especially those involving bacteria, such as As(V) reduction, As(III) oxidation and the various methylation reactions. These reactions protect bacteria from the toxic effects of arsenic (by enhancing the resistance mechanisms), and in the case of some species, contribute to energy metabolism processes [25].

Two major biological mechanisms can describe the reduction of arsenate into arsenite. The first one is related to the detoxification of the cells. Arsenate ions enter the cells via phosphate transporters due to structural homologies with phosphate ions. After reaching the cytoplasm, As(V) is reduced into As(III) by the enzyme *ArsC* before being excreted from the cell [26]. The second process, known as dissimilatory reduction, has been described in bacteria belonging to various phylogenetic groups including obligate or facultative anaerobic micro-organisms [27]. In this case, bacterial cells gain metabolic energy by "breathing" arsenic (known as respiratory arsenate reductase), since they use As(V) as an electron acceptor [28]. These two systems play a non-negligible role in the solubilisation of arsenic leading to major contaminations of aquatic environments.

A wide range of bacteria isolated from various contaminated environments were described for their ability to synthesize arsenite and thus to oxidize As(III) enzymatically. They include heterotrophic bacteria as well as chemoautotrophic bacteria in which As(III) serve as an electron donor reducing oxygen or nitrate. In this latter case, the energy produced is used to fix CO_2, which provide bacteria with the carbon required for growth. In populations where both heterotrophic and chemoautotrophic bacteria are present, heterotrophic As(III)-oxidizing bacteria can develop using organic substances synthesized by chemoautotrophic bacteria [29]. The biological oxidation of As(III) to As(V) by iron and manganese oxidizing bacteria has been reported and it was also confirmed that trivalent arsenic can be efficiently treated without any additional use of chemicals in this bioprocess [30]. The bioconversion of arsenite can also be accomplished via methylation or oxidation-reduction of the arsenic species found in the water. Their bioconversion efficiency can be increased by genetic modification of the intervening microorganisms [31].

The study of microbial biofilms metal dynamics in acid rock drainage provided evidence of the stable accumulation of metals in these bioreactors, in which secrete polymers are able to immobilize metal compounds by passive sequestration processes, avoiding its use to develop bioremediation methods. In the presence of arsenic, bacteria such as *H. arsenicoxydans* produce large amounts of exopolymers that can be used to detoxify natural waters contaminated with arsenic [32].

Arsenic mobility in aquifers is determined by complex interactions of hydrology, water chemistry, and biological processes. Anaerobic microorganisms play an important role in the release of arsenic in iron-rich aquifers, where arsenic is typically sequestered by iron minerals. Since As(V) reduction is chemically slow, accelerated reduction in natural systems is often achieved biologically [33]. The observed transient release is especially relevant to dynamic (flowing) systems, where release of As(III) has been observed in the presence of bacterial iron and arsenic reduction [34]. Bacterial sulphate reduction is recognised as a mechanism for sequestering metals in contaminated environments, primarily through precipitation of metal sulphides. The sulphide produced by bacteria can reduce As(V) and precipitate As(III) in sulphide phases [35].

New developments in the biotechnological field have focused on phyto-remediation and biofiltration processes [36,37] that have revealed to be efficient and environmentally sustainable. In phyto-remediation, plant and fungal biomass is used as a renewable adsorbing material in a process that is similar to chemical compound removal. The adsorbing capacity of this biomass is superior to that of activated carbon and of some synthetic resins used in water treatment.

A few of the most promising emerging technologies use UV radiation, solar radiation and/or biological processes (bacterial or plant action) for arsenite removal. For example, the conversion of arsenite surpasses 95% when UV or solar radiation is used in the presence of iron or sulphites [38].

Table 2 presents a comparative synthesis of the main emerging techniques for arsenic removal in water treatment (adapted from [12]).

Table 2. Comparison of the main emerging techniques for arsenic removal.

Oxidation process	Parameters	Removal efficiency (%)	Advantages	Disadvantages
UV Ray and sun light in presence of iron	As_0 = 10 mg/L pH – 1.5 3 Fe_0 = 180 mg/L Fe(II)/As(III) = 24	>96	No use of chemicals	pH is lower
UV Ray in presence of sulphite	As_0 = 0.47 mg/L pH = 9 air bubble supplied	>96	Effluent amenable to all techniques for As(V) removal	Sulphite content influences the efficiency
Iron-oxidizing bacteria in presence of Fe and Mn	As_0 = 35.60 ppb pH = 7.2 Fe_0 = 2.8 mg/L, Mn_0 = 0.6 mg/L	>80	No use of chemical; cheaper and eco-friendly; indigenous microbes	Not well established
Sulphate-reducing bacteria	As_0 = 10 mg/L pH = 4.5–7 Fe_0 = 20 mg/L	>78	No use of chemical; cheaper and eco-friendly; indigenous microbes	Not well established

3.4. Alternative Technologies

Some alternative safe water options applied in West Bengal and Bangladesh include clay filters, deep tube wells, dug wells, surface and rainwater harvesting and solar distillation. Solar distillation techniques use the sun's energy to evaporate water, which is then re-condensed. This process of evaporation and re-condensation separates all chemicals, including arsenic, from the water [16].

The SORAS (*Solar Oxidation and Removal of Arsenic*) process has been used in the rural Andes regions in Latin America. It is based on the adsorption of As(V) onto iron oxides and hydroxides using UV radiation [33,34,35] and the addition of citrate as a catalyst for the formation of oxidising radicals that allow the conversion of arsenite to arsenate (Figure 2).

Figure 2. Arsenic removal using the SORAS process.

In this context, the combination of solar oxidation with one of the conventional adsorption processes might be a pertinent development towards yet another promissory alternative to arsenic removal. The goal would be to optimise the SORAS process by incorporating complementary conventional techniques, thereby increasing its reliability [12]. To that effect, it was proposed that a UV radiation measurement cell be programmed so that the water to treat is automatically re-routed to a chemical oxidation process, whenever the available UV radiation is less than that required for arsenite oxidation. This alternative technology appears as particularly suitable for small and medium-size water supply systems. By integrating a solar radiation (renewable energy source) oxidation technology with conventional reactive filtration or adsorption processes, it is also aiming for the reduction of arsenic removal costs.

4. CASE STUDY: REHABILITATION OF THE WTP OF MANTEIGAS

4.1. Problem and Scope Definition

The municipality of Manteigas is served by the public water supply to the Manteigas sub-system, whose intake is located in springs from the Portuguese Estrela Mountain massif (Figure 3). The hydrogeological environment features mineralisation outcroppings that are rich in arsenopyrite generally associated to phyllite, meta-greywacke and quartzite. Consequently, raw water from the Paulo Luís Martins spring—the main source supplying the 2002-rehabilitated WTP of Manteigas—shows an annual average arsenic concentration of 12 ppb, a value that is slightly above the legal limit of 10 ppb.

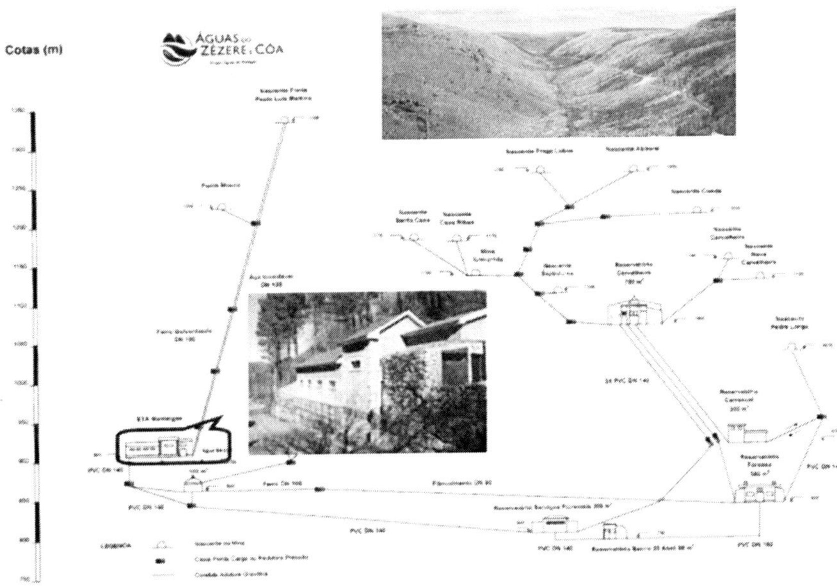

Figure 3. Manteigas water supply sub-system and the location of its WTP.

The AdZC Company has been conducting analytical monitoring of the influent raw water at the WTP since 2002. Their monitoring reports confirm the expected seasonal variation of arsenic concentrations between 9 and 18 ppb. Lower values are observed during dry seasons and higher values correspond to periods of greater precipitation [5].

At the time of the WTP rehabilitation, it was not possible to implement a sustainable solution for addressing the non-compliance issue, since the options available were not able of adjusting to the limitations and specificity of the small water supply system, including the need to remove very small quantities of arsenic. Limited space availability and the sub-system's small service size—approximately 3,000 inhabitants—compromised the viability and the economic sustainability of accommodating a more complex solution for arsenic removal.

Aside from equipment cost, the WTP would require expansion onto adjacent land, which would imply additional expropriation costs. Such an investment would compromise the balance of the existing economic model, leading to an increase in tariffs that would eventually become unfeasible.

4.2. Methodology and Laboratorial-Scale Results

The AdZC Company promoted internal research work aiming to develop an effective arsenic removal process that would suit the characteristics and size of the sub-system. The methodology included two distinct phases. Phase 1 consisted of laboratory-scale experimental trials that were carried-out in order to evaluate the removal efficiencies of pre-selected conventional techniques. Results from these trials allowed the later testing—phase 2—of the most efficient alternatives at full-scale, taking into account the real hydraulic conditions at the WTP.

The pre-selected techniques were based on a scheme consisting of oxidation using chlorine dioxide followed by coagulation with aluminium sulphate and sand filtration. The results of the first set of experimental trials (Figure 4) produced removal efficiencies above 37% for raw water exhibiting an initial concentration of 10 ppb. These removal efficiencies were independent of the coagulant dosage in the experiments.

A second set of laboratory trials was conducted using a raw water sample with an initial concentration of 18 ppb. Removal efficiencies between 77 and 88% were attained through a combination of "oxidation + filtration" or "oxidation + coagulation + filtration" processes. Table 3 summarises the six different laboratorial procedures and the obtained results.

Figure 4. Laboratory experimental trials for the selection of the most effective and viable methods.

Table 3. Results of the second stage of laboratory trials for arsenic removal.

Sampling	Process description	Arsenic (ppb)
Raw water		18
1	Oxidation with sodium hypochlorite (0.33 mg/L) Sand filtration	2
2	Oxidation with sodium hypochlorite (0.33 mg/L) Coagulation with aluminium salt (10 ppm) Sand filtration	4
3	Oxidation with sodium hypochlorite (0.33 mg/L) Coagulation with aluminium salt (20 ppm) Sand filtration	4
4	Oxidation with chlorine dioxide (0.38 mg/L) Sand filtration	2
5	Oxidation with chlorine dioxide (0.38 mg/L) Coagulation with aluminium salt (10 ppm) Sand filtration	3
6	Oxidation with chlorine dioxide (0.38 mg/L) Coagulation with aluminium salt (20 ppm) Sand filtration	3

Note that at this scale the hydraulic conditions tested did not match the full-scale conditions observed at the WTP. Based on the laboratorial results, a pilot-installation was tested at the WTP. It consisted of a double-media pressure filter (0.20 m sand and 0.05 m granulated coal) preceded by pre-oxidation with sodium hypochlorite. The first set of results was not satisfactory because the concentration of arsenic in the filter effluent was a non-compliant, albeit slightly, 11 ppb (39% removal).

In order for such a treatment scheme to work, the WTP would have to be fitted with a pre-oxidation process followed by coagulation and filtration steps, which would result in a heavy and costly construction intervention. The results obtained did not support this option and thus, it became necessary to ponder other treatment alternatives, particularly those of easier implementation.

4.3. Real-Scale Implementation and Results

An emerging arsenic removal technology—reactive filtration process based on arsenic adsorption onto iron oxide—was selected for testing. A series of laboratory trials was conducted to simulate the operation of a sand and iron oxide filter, using raw water presenting an arsenic concentration of 12 ppb. The final results showed arsenic levels in the filtered water to be below 1 ppb, achieving very high removal efficiency (near 95%). These quite satisfactory results demonstrated that the problem could be solved through a viable, effective, simple, economical and rather simple system.

Given the high removal efficiency obtained during the experimental trials, a filter was installed at the WTP using iron oxide adsorbent material designed to treat only a fraction (10 $m^3 \times h^{-1}$) of the maximum flow rate (35 $m^3 \times h^{-1}$) required by this water supply network. This ensured the advantageous possibility of retrofitting onto existing tanks with no need for additional space occupation (Figure 5).

Figure 5. Installation of the reactive filter using iron oxide adsorbent medium.

The combination of filtered and unfiltered flows (25 $m^3 \times h^{-1}$) allowed for a final water arsenic concentration below 8 ppb, as depicted in Figure 6.

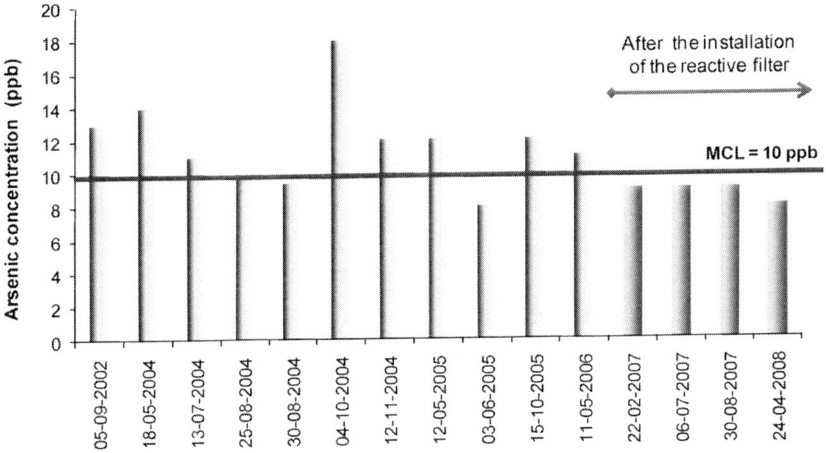

Figure 6. Arsenic concentration in the treated water of the Manteigas WTP.

The adequacy of this solution lies also in its adaptability to future legal requirements. It is conceivable that the flexibility of operation could be easily adapted to comply with lower arsenic levels, should they be imposed in the future. As designed, a simple adjustment to the filtered/unfiltered water ratio could mean continuing compliance with future more restrictive drinking water quality standards.

5. CONCLUSIONS

The presence and levels of arsenic in natural water sources must be periodically controlled, since the mechanisms that foster the prevalence of high concentrations are many and difficult to identify. However, by identifying risk areas, that is, regions where natural waters normally present high arsenic concentrations, it is possible to promote the establishment of arsenic hazardous criteria to feed Water Safety Plan risk analysis and risk management approaches.

Treating arsenic-polluted dinking water is vital as an instrument of public health prevention policies, since acute and chronic exposure to arsenic can cause a variety of diseases, including fatal cancers. Unfortunately, the full extent of arsenic-related health problems has not yet been fully identified and quantified.

Available arsenic removal technologies are generally complex, expensive and often inadequate for As(III). Therefore, it is of paramount importance that adequate and sustainable treatment options are selected in order to comply with increasingly more stringent quality standards. Ideally, these should be supported by systematic epidemiological studies that are all encompassing and conclusive.

The progressive reduction in the number of water sources suitable for public supply, the increasingly restrictive nature of quality standards for drinking water, and the potential impacts of climate change in the quantities of potable water available, make the need for alternative, sustainable and low-cost arsenic removal technologies all the more pertinent. In this context, the utilisation of new adsorbing materials and bio-oxidation techniques take on particular relevance.

The solution for the financial and operational constraints usually endured by managing entities of small and medium-size public water supply systems must include the adoption of arsenic removal methodologies based on a combination of conventional, alternative and emerging processes. Not only should these aim for significant reductions of operation costs, they should also help prevent the rash acquisition of expensive and pre-formatted solutions that may not always allow the control and deep knowledge of removal mechanisms.

The results obtained demonstrate the successful implementation of an effective and economically sustainable solution for correcting arsenic levels in raw water with excellent microbiological quality. The treatment option selected used technology based on a reactive filtration process that uses iron oxide as an adsorbent and that allowed for very high removal efficiencies (over 95%).

The case described in this research work constitutes a good example of the research effort initiated and developed by a water supplier that, using the available in-house human resources and expertise, was able to implement a rather effective and low-cost solution at the WTP of Manteigas.

REFERENCES

1. Nordstrom, D.K. Worldwide occurrences of arsenic in groundwater. *Science* **2002**, *296*, 2143–2145.
2. Yoshida, T.; Yamanchi, H.; Jun, G.F. Chronic health effect in people exposed to arsenic via the drinking water: dose-response

relationship in review. *Toxicol. Appl. Pharmacol.* **2004**, *198*, 243–252.

3. Lamm, S.H.; Engel, A.; Kruse, M.B.; Feinleib, M.; Byrd, D.M.; Lai, S.; Wilson, R. Arsenic in drinking water and bladder cancer mortality in the US: an analysis based on 133 US countries and thirty years of observation. *J. Occup. Environ. Med.* **2004**, *46*, 298–306.
4. *Guidelines for Drinking Water Quality*; World Health Organisation: Geneva, Switzerland, 1993; p. 41.
5. Cardoso, S.J.A.; Alçada, A.J. Correcção dos níveis de arsénio na ETA de Manteigas. In Proceedings of 13th ENaSB—Conference on Sanitary and Environmental Engineering, Lisbon, Portugal, October 14–17, 2008.
6. Ávila, P.; Pereira, M.R.; Novais, H.; Ferreira, A. Arsénio nos solos, sedimentos e águas na envolvente da mina do Tuela (Vinhais, NE Portugal). In Proceedings of VII National Congress of Geology, Lisbon, Portugal, June 29–July 13, 2006.
7. Oremland, S.; Stolz, J.F. Ecology of arsenic. *Science* **2003**, *300*, 939–944.
8. Smedley, P.L.; Kinniburgh, D.G. A review of the source, behaviour and distribution of arsenic in natural waters. *Appl. Geochem.* **2002**, *17*, 517–568.
9. Duarte, A.A.L.S.; Cardoso, S.J.A.; Alçada, A.J. Remoção de arsénio em sistemas de abastecimento de água. Um caso de estudo. *Águas Resíduos* **2009**, *9*, 28–37.
10. Ratnaike, R.N. Acute and chronic arsenic toxicity. *Postgrad. Med. J.* **2003**, *79*, 391–396.
11. *Draft Final Guidance for Carcinogen Risk Assessment Forum*; US Environmental Protection Agency (USEPA): Washington, DC, USA, 2003.
12. Cunha, P.D.R.; Duarte, A.L.S. Remoção de arsénio em águas para consumo humano. In Proceedings of 13th ENaSB—Conference on Sanitary and Environmental Engineering, Lisbon, Portugal, October 14–17, 2008.
13. Mondal, P.; Majumder, C.B.; Mohantya, B. Laboratory based approaches for arsenic remediation from contaminated water: recent developments. *J. Hazar. Mater.* **2006**, *137*, 464–479.

14. Katsoyiannis, I.A.; Hugh, S.J.; Ammann, A.; Zikoudi, A.; Hatziliontos, C. Arsenic speciation and uranium concentrations in drinking water supply wells in Northern Greece: correlations with redox indicative parameters and implications for groundwater treatment. *Sci. Total Environ.* **2007**,*383*, 128–140.

15. Johnston, R.; Heijnen, H. Safe water technology for arsenic removal. In *Technologies for Arsenic Removal from Drinking Water*; Ahmed, M.F., Ali, M.A., Adeel, Z., Eds.; Bangladesh University of Engineering and Technology: Dhaka, Bangladesh, 2001.

16. National Drinking Water Advisory Council (NDWAC). *Final Report of the Arsenic Cost Working Group to the National Drinking Water Advisory Council*; USEPA: Washington, DC, USA, 2001.

17. Hao, J.M.; Han, M.J.; Meng, X.G. Preparation and evaluation of thiol-functionalized activated alumina for arsenite removal from water. *J. Hazar. Mater.* **2009**, *167*, 1215–1221.

18. Siegel, M.; McConnell, P.; Ilges, A.; Chen, H.-W.; Ghassemi, A.; Thompson, R. Development and evaluation of innovative arsenic adsorption technologies for drinking water by the arsenic water technology partnership. In Proceedings of the National Groundwater Association Meeting on Naturally Occurring Contaminants Conference, Albuquerque, NM, USA, February 6–7, 2006.

19. Mohan, D.; Pittman, C.U., Jr. Arsenic removal from water/wastewater using adsorbents—a critical review. *J. Hazard. Mater.* **2007**, *142*, 1–53.

20. Pena, M.E.; Korfiatis, G.P.; Patel, M.; Lippincott, L.; Meng, X.G. Adsorption of As(V) and As(III) by nanocrystalline titanium dioxide. *Water Res.* **2005**, *39*, 2327–2337.

21. Boddu, V.M.; Abburi, K.; Talbott, J.L.; Smith, E.D.; Haasch, R. Removal of As(III) and As(V) from aqueous medium using chitosan-coated biosorbent. *Water Res.* **2008**, *42*, 633–642.

22. Fostier, A.H.; Pereira, M.D.S.S.; Rath, S.; Guimarães, J.R. Arsenic removal from water employing heterogeneous photocatalysis with TiO_2 immobilized in PET bottles. *Chemosphere* **2008**, *72*, 319–324.

23. Mendes, G.; Bellato, C.R.; Marques, J.D. Heterogeneous photocatalysis with TiO_2 for the oxidation of arsenic and its removal from water by co-precipitation with ferric sulphate. *Quím. Nova* **2009**,*32*, 1471–1476.

24. Pokhrela, D.; Viraraghavan, T. Biological filtration for removal of arsenic from drinking water. *J. Environ. Manage.* **2009**, *90*, 1956–1961.

25. Lièvremont, D.; Bertin, P.N.; Lett, M.C. Arsenic in contaminated waters: biogeochemical cycle, microbial metabolism and biotreatment processes. *Biochimie* **2009**, *91*, 1229–1237.

26. Mukhopadhyay, R.; Rosen, B.P. Arsenate reductases in prokaryotes and eukaryotes. *Environ. Health Perspect.* **2002**, *110*, 745–748.

27. Paez-Espino, D.; Tamames, J.; de Lorenzo, V.; Canovas, D. Microbial responses to environmental arsenic. *BioMetals* **2009**, *22*, 117–130.

28. Macy, J.M.; Santini, J.M.; B.V.; Pauling, B.V.; O'Neill, A.H.; Sly, L.I. Two new arsenate/sulphate-reducing bacteria: mechanisms of arsenate reduction. *Arch. Microbiol.* **2000**, *173*, 49–57.

29. Battaglia-Brunet, F.; Dictor, M.C.; Garrido, F. An arsenic(III)-oxidizing bacterial population: selection, characterization, and performance in reactors. *J. Appl. Microbiol.* **2002**, *93*, 656–667.

30. Casiot, C.; Morin, G.; Jullot, F.; Burneel, O.; Personne, J.C.; Leblanc, M.; Duquesne, K.; Bonnefoy, V.; Poulichet, F.E. Bacterial immobilization and oxidation of arsenic in acid mine drainage (Carnoules Creek, France). *Water Res.* **2003**, *37*, 2929–2936.

31. Kostal, J.; Yang, R.; Nu, C.H.; Mulchandani, A.; Chen, W. Enhanced arsenic accumulation in engineered bacterial cells expressing ArsR. *Appl. Environ. Microbiol.* **2004**, *70*, 4582–4587.

32. Muller, D.; Médigue, C.; Koechler, S.; Barbe, V.; Barakat, M.; Talla, E.; Bonnefoy, V.; Krin, E.; Arsène-Ploetze, F.; Carapito, C.; Chandler, M.; Cournoyer, B.; Cruveiller, S.; Dossat, C.; Duval, S.; Heymann, M.; Leize, E.; Lieutaud, A.; Lièvremont, D.; Makita, Y.; Mangenot, S.; Nitschke, W.; Ortet, P.; Perdrial, N.; Schoepp, B.; Siguier, P.; Simeonova1, D.D.; Rouy, Z.; Segurens, B.; Turlin, E.; Vallenet, D.; van Dorsselaer, A.; Weiss, S.; Weissenbach, J.; Lett, M.C.; Danchin, A.; Bertin, P.N. A tale of two oxidation states: bacterial colonization of arsenic-rich environments. *PLoS Genet.* **2007**,*3*, 518–530.

33. Saalfield, S.L.; Bostick, B.C. Changes in iron, sulfur, and arsenic speciation associated with bacterial sulfate reduction in ferrihydrite-rich systems. *Environ. Sci. Technol.* **2009**, *43*, 8787–8793.

34. Herbel, M.; Fendorf, S. Transformation and transport of arsenic within ferric hydroxide coated sands upon dissimilatory reducing bacterial activity. *ACS Symp. Ser.* **2005**, *915*, 77–90.

35. Kaksonen, A.H.; Puhakka, J.A. Sulfate reduction based bioprocesses for the treatment of acid mine drainage and the recovery of metals. *Eng. Life Sci.* **2007**, *7*, 541–564.

36. Katsoyiannis, I.A.; Zouboulis, A.I. Application of biological processes for the removal of arsenic from groundwater. *Water Res.* **2004**, *38*, 17–26.

37. Murugesan, G.S.; Sathishkumar, M.; Swaminathan, K. Arsenic removal from groundwater by pre-treated waste tea fungal biomass. *Bioresource Technol.* **2006**, *97*, 483–487.

38. Zaw, M.; Emett, M.T. Arsenic removal from water using advanced oxidation processes. *Toxicol. Letter* **2002**, *133*, 113–118.

39. Wegelin, M.; Getcher, D.; Hug, S.; Mahmud, A. SORAS—a simple arsenic removal process. In Proceedings of 26th WEDC Conference—Water, Sanitation and Hygiene: Challenges of the Millennium, Dhaka, Bangladesh, November 5–9, 2000.

40. Meichtry, J.M.; Lin, H.J.; Fuente, L.; Levy, I.K.; Gautier, E.A.; Blesa, M.A.; Litter, M. Low-cost TiO_2photocatalytic technology for water potabilization in plastic bottles for isolated regions. *J. Sol. Energ-T ASME* **2007**, *129*, 119–126.

41. Zhang, F.S.; Itoh, H. Photocatalytic oxidation and removal of arsenite from water using slag-iron oxide-TiO_2 adsorbent. *Chemosphere* **2006**, *65*, 125–131.

10
CHAPTER

Review on Optimization of Drinking Water Treatment Process

Mohamed Farhaoui[1,2], Mustapha Derraz[2]

[1]National Office of Electricity and Drinking Water, Meknes, Morocco

[2]Biology Department, University of Moulay Ismail, Meknes, Morocco

ABSTRACT

In the drinking water treatment processes, the optimization of the treatment is an issue of particular concern. In general, the process consists of many units as settling, coagulation, flocculation, sedimentation, filtration and disinfection. The optimization of the process consists of some measures to decrease the managing and monitoring expenses and improve the quality of the produced water. The objective of this study is to provide water treatment operators with methods and practices that enable to attain the most effective use of the facility and in consequence optimize the cubic meter price of the treated water. This paper proposes a review on optimization of drinking water treatment process by analyzing all of the water treatment units and gives some solutions in order to maximize the water treatment performances without compromising the water quality standards. Some practical solutions and methods are performed in the water treatment plant located in the middle of Morocco (Meknes).

Keywords: Optimization, Coagulation Process, Turbidity Removal, Water Treatment

1. INTRODUCTION

The demand on water supply is increasing over the last century due to improved lifestyle, industrial development and population growth. This increased demand is facing a paradox to produce treated water with high quality at lower cost. In order to reduce the water cost, it is very important to optimize the operating expenses in the water treatment plant (power, chemicals, operator's expenses...) and many measures should be performed in this vision.

The optimization of water treatment plant is not a disciplinary to maximize the treatment objective and minimize the cost of the water produced. But it consists of understanding the functioning of the treatment plant and treasure the experience of the operators in dealing with all treatment process related to different aspects.

This paper addresses the problem of optimizing of the conventional drinking water plant and proposes some practical to reach this goal. This paper is organized as fellows. After an introduction of the objective of this study, the water treatment operation is described in Section 2. Review on optimization of conventional drinking water treatment plant is discussed in Section 3. In Section 4, some practical methods are explained.

2. WATER TREATMENT OPERATION

Meknes region is located in the middle of Moroccan Kingdom, has a Mediterranean climate with continental influences. The temperature shifts from cool and cold in winter to hot days in the summer months of June - September. The agriculture and industry are the mean activities in the region.

This study was developed in a water treatment plant located in Meknes, whose source is two big springs Bittit (630 l/s) and Ribaa (400 l/s). The quality of water, produced by the spring, changes according to the rainfall in the region. Sometimes, it can be affected by the snow in the Atlas Mountains. The treatment water plant, as part of other water resources, water to more than 700,000 inhabitants of Meknes city, and it has a nominal capacity of 600 l/s of treated water. This water treatment plant is chosen regarding to the variation of the raw water turbidity and because of the availability of data. Figure 1 presents a schematic overview of the various operations necessary to treat the water.

Many measurements of variables recorded by streaming current detectors such as: turbidity level, PH, conductivity, temperature is needed to carry out

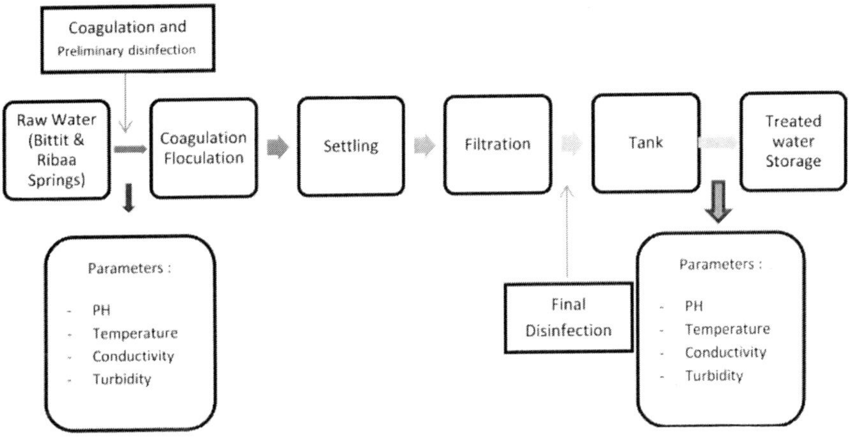

Figure 1. Simplified synopsis of the water treatment plant.

Table 1. Statistical summary of raw water conditions from 01/01/2013 to 31/12/2015 (ONEE, 2015).

Variables	Min	Max
Turbidity: Bittit (NTU)	1.7	850
Turbidity: Ribaa (NTU)	1.62	960
PH	6.80	7.74
Temperature: (°C)	14	24.70
Conductivity micro s/cm	509	624

the jars test in order to determine the optimal dose of the aluminum sulfate. Table 1 shows the raw water variables used in this study, and they present the following variation intervals.

The variation of PH, conductivity and temperature from 01/01/2013 to 31/12/2015 is given by Figures 2(a)-(c).

In the rainfall period, the turbidity of raw water changes from time to time as shown in the graph below, the turbidity of the raw water can increase to reach levels more than 500 NTU as shown inFigure 3.

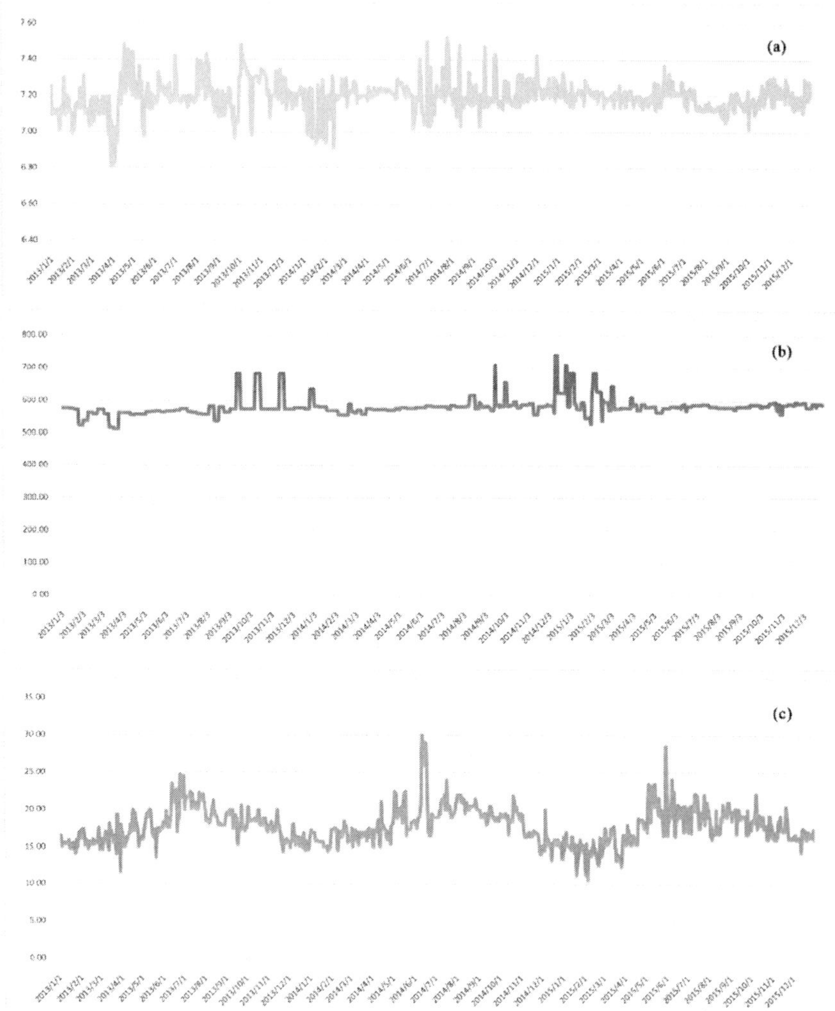

Figure 2.Evolution of PH. (a) conductivity in micro S/cm; (b) and temperature in °C; (c) of raw water from 01/01/2013 to 31/12/2015.

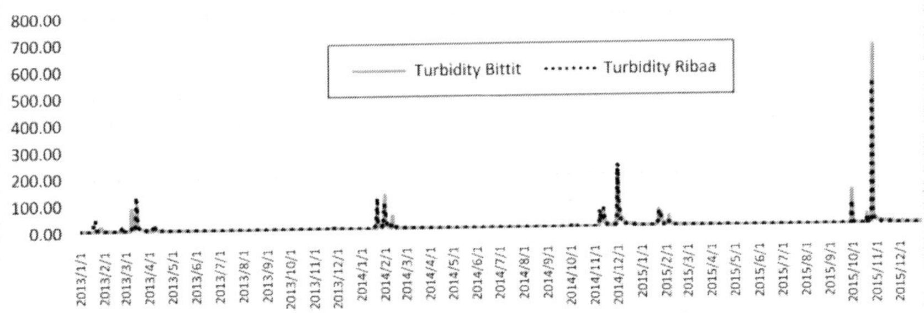

Figure 3. Statistical data of turbidity level of the spring's water from 01/01/2013 to 31/12/2015 (ONEE, 2015).

Table 2. Turbidity levels distribution from 2009 to 2015 (number of days per turbidity level).

Year	**Number of days**					**Total**
	Turbidity less or equal than 5 NTU	Turbidity more than 5and less or equal than 10 NTU	Turbidity more than 10 and less or equal than 20 NTU	Turbidity more than 20 and less or equal than 40 NTU	Turbidity more than 40 NTU	
2009	147	60	101	34	23	365
2010	0	0	113	148	104	365
2011	0	132	144	59	30	365
2012	301	38	17	5	5	366
2013	260	74	23	8	0	365
2014	247	62	32	20	4	365
2015	184	132	31	10	5	365
Total	1139	498	464	284	171	2556
Aggregated data	1139	1637	2101	2385	2556	
Percentage	**45%**	**19%**	**18%**	**11%**	**7%**	
Percentage of Aggregated data	**45%**	**64%**	**82%**	**93%**	**100%**	

However, the turbidity level is less than 10 NTU this three last years (2013, 2014 and 2015) for more than 88% of the year and more than 64% of the year (1637/2556 days); the turbidity is less than 10 NTU for the six last years as shown by Table 2.

Table 3 gives the max and min of raw water turbidity by month from 2013 to 2015.

The chemicals used in the water treatment process consume about 50% the total operating expenses of the water treatment. The energy cost is between 10% and 15% related to the total cost in 2013, 2014 and 2015. Figure

4 gives the operations expenses of the water treatment plant in the three last years.

In addition, used as coagulant, the aluminum sulfate (SA) consumption is more than 65% of the total chemicals consumption in the water treatment plant. Le polyelectrolyte (Poly) consumption is less than 10% and the chlorine is between 16% and 26% of the total chemicals used in the water treatment plant according to the water quality as shown by Figure 5.

3. REVIEW ON OPTIMIZATION OF DRINKING WATER TREATMENT PLANT

In order to remove turbidity and contaminants from raw water, the treatment comprises many steps as the settling, aeration, coagulation, sedimentation, filtration and disinfection. The optimization of water treatment plant concerns the different steps of the process. The main goals are to:

- Achieve the chemical cost performance.
- Manage and control the energy consumption.
- Reduce the water wastes.

Table 3. Statistical data of turbidity min and max (NTU) measured in 2013, 2014 and 2015 per month (ONEE, 2015).

Year: 2013				
	Bittit Spring		Ribaa Spring	
Month	min	max	min	max
January	3.70	18.95	10.00	50.00
February	4.50	19.00	3.95	20.65
March	4.95	98.40	4.50	136.66
April	6.40	21.95	4.90	32.77
May	3.93	6.30	3.80	5.30
June	3.72	4.40	3.34	4.89
July	3.30	4.33	2.92	3.62
August	3.00	3.80	2.70	3.46
September	3.10	4.64	2.68	4.15
October	2.90	3.99	2.46	3.62
November	2.70	7.44	2.90	10.00
December	3.07	4.23	2.60	3.95

Year: 2014	Bittit Spring		Ribaa Spring	
	min	max	min	max
January	3.06	81.69	3.00	120.66
February	6.48	128.00	5.93	78.90
March	4.86	6.91	3.90	5.65
April	4.30	4.90	3.30	4.19
May	3.17	4.77	2,48	3.37
June	3.00	3.86	2.60	2.87
July	2.36	4.50	2.16	2.49
August	1.74	4.20	1.88	2.28
September	1.80	3.50	1.64	1.90
October	1.89	4.20	1.67	11.64
November	1.80	55.50	1.68	69.29
December	6.57	197.50	5.06	235.89

Year: 2015	Bittit Spring		Ribaa Spring	
	min	max	min	max
January	5.09	63.10	3.96	61.59
February	6.33	38.52	3.98	19.45
March	4.97	6.27	3.26	4.64
April	3.65	5.60	2.80	3.76
May	3.50	5.23	2.68	3.06
June	3.40	4.52	2.61	3.19
July	2.80	4.75	2.40	3.60
August	2.60	3.26	2.45	2.74
September	2.70	128.45	2.35	86.63
October	4.35	669.71	3.22	524.03
November	6.76	17.18	3.99	13.19
December	7.18	9.54	3.42	4.02

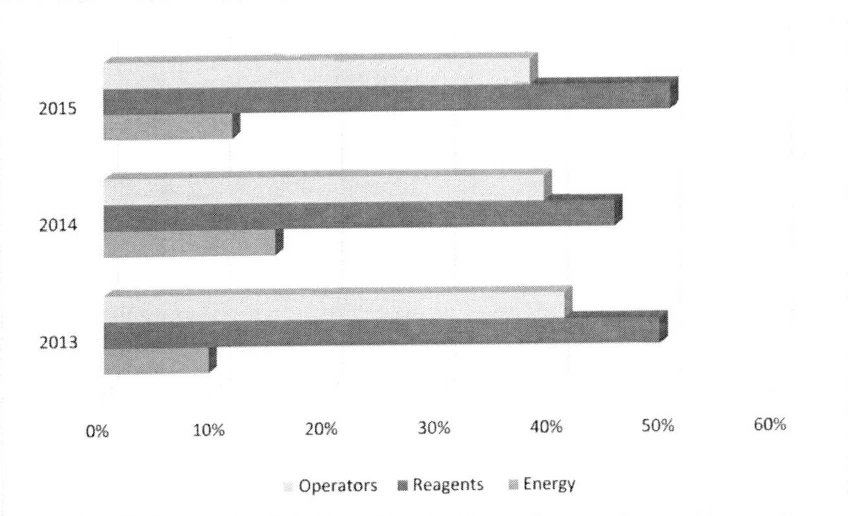

Figure 4. Operations expenses of the water treatment plant in 2013, 2014 and 2015 (ONEE, 2015).

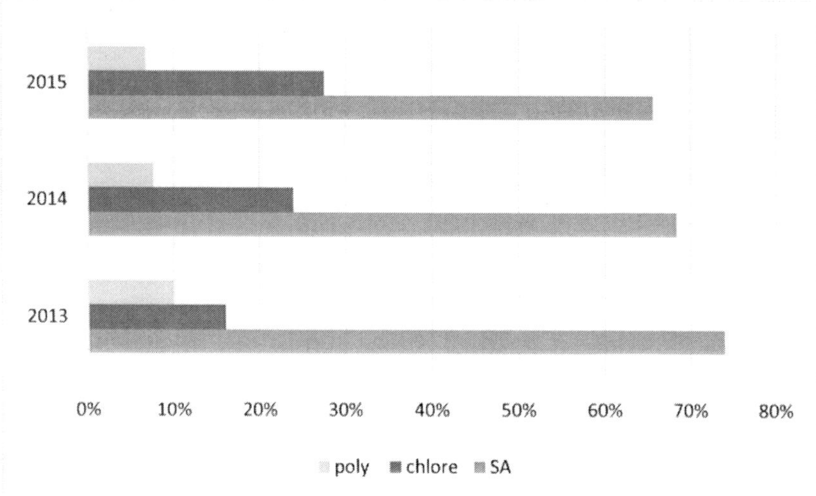

Figure 5. Percentage of chemicals expenses consumed by the water treatment plant in 2013, 2014 and 2015 (ONEE, 2015).

Many research and studies are performed in order to increase the water treatment plant performances. In this section, one aspect related to the optimization is treated which is the coagulant consumption.

3.1. Coagulant and Coagulant Aid

In the water treatment plant, colloidal particles are separated by means of a chemical coagulation process: consisting in the charge destabilization of the suspended particles by adding coagulant. The coagulant used is aluminum sulfate; it is the most widely used coagulant in Morocco as well as many other countries in the drinking water industry. It is mainly used because of its effectiveness, accessibility and low price. As a common practice, aluminum sulfate is applied according to the jars test results. The main difficulty is how to optimize the aluminum sulfate dosage related to raw water characteristics by using other cheaper products. Some attempts have been made to improve the effectiveness of the aluminum sulfate or to substitute this coagulant by another natural, available and cheaper. Finding of various coagulation processes have been reported in literature. Some of these include; studying the effect of using the bentonite on the coagulation in the treatment of low turbidity [1] . Also, Mukheled (2012) [2] used Date seeds and Pollen Sheath as coagulant to treat different levels of turbidity

(75, 150 and 300 NTU). Eman et al. (2010) [3] tried Moringa Oleifera seed in the coagulation process to treat low turbid water in Malaysia. However, Aho et al. (2012) [4] highly recommended the use of this natural coagulant in the domestic turbid water purification in Nigeria. Other natural coagulants are proposed as an important alternative in the water treatment plant. The coagulants are from plant origin such as nirmali seed and maize [5] , cassia angustifolia seed [6] , mesquite bean and cactus latifaria [7] , chestnut and acorn [8] , Cocciniaindica fruit mucilage [9] and from different leguminous species [10] . Also, the Bhindi seed, Methi, Beheda, Guar seeds and Drum stick can be used as coagulant in 100 to 1200 NTU turbid water range with remarkable percentage removal from 70% to 93% [11] . In the other hand, Cocciniaindica, Strychnospotatorum and Cactus opuntia were used as natural coagulant and improved the quality of the filtered water [12] . Those natural products have coagulating activity in the treatment of turbid water and can be used as coagulant or as coagulant aid with other synthetic and industrial coagulants (aluminum sulfate...) in order to reduce the coagulant consumption in the water treatment plant [13] .

3.2. Coagulant Control and Monitoring

As a common practice, aluminum sulfate is applied according to the jars test results. The problem is to determine the optimal dose of aluminum sulfate related to raw water characteristics. Both manual and automatic methods are used to predict optimum coagulant dose [14] [15] . Automatic method is ensured by streaming current detectors [16] . However, manual method is consisting to determine the quantity of the coagulant to apply experimentally and based on the jar test results. Jar test involves taking a raw water samples and applying different quantities of coagulant to each sample [15] . After a short period of time, each sample is assessed for water quality and the dosage that produces the optimal result used a set point. This operation should be repeated by the operators each time when the quality of raw water changes. The aluminum sulfate is the compound likely to be mathematically modeled and therefore its value can be estimated according to the data available in the treatment plant. The optimization of using the coagulant is very interesting approach because under dosing of coagulant can lead to poor quality drinking water while too much coagulant leads to many operating problems (less efficient filtration and sedimentation, PH), healthy problems and can increase the cost of treated water.

Some attempts have been made to model the relationships between raw water quality characteristics and the optima coagulant dosage rate.

4. PRACTICAL METHODS OF WATER TREATMENT PLANT OPTIMIZATION

4.1. Understanding of the Water Treatment Process

The understanding of the water treatment functioning is a very important step on the performances optimization program. A systematic gathering of information about plant performances are need such as:

- Data trending and analysis.
- Check plant design criteria against actual and the controlling operations to improve the plant performances.
- Track chemical dosing and use indicators.
- Field measurements and visual observation.
- Collaboration of operators (quality water operators, maintenance team and managing operator).
- Plant specific operating guidelines and procedures.
- Accuracy and control of the on-line measurement devices.

4.2. The Aluminum Sulfate Consumption

The sludge produced in the settling step of the treatment process is used not only to reduce the aluminum sulfate dose and the sludge volume but also in order to improve the water quality.

A standard jar test apparatus equipped with six paddles rotating in a set of six beakers is used to simulate coagulation, flocculation and sedimentation processes. At the first, Control experiments for coagulation tests are performed in order to determine the optimal dose of the aluminum sulfate in normal conditions. The selected level of turbid water (1 L) is filled into the beakers and various doses in the range from 10 to 100 mg/l of sludge and the aluminum sulfate according to the results of the first jar test determining the optimal dose of the inorganic coagulant in normal conditions are separately added in the beakers and mixed rapidly (300 rpm) for one minute. The mixing speed was then reduced to 40 rpm for 20 minutes. Then the stirrer is turned off and the suspensions are allowed to settle for different periods of time ranging from 30 to 120 minutes under quiescent conditions. After each period of settling time, supernatant samples of each beaker in the jar test is withdrawn from located 10 cm below the water level and residual turbidity is measured.

Standard Jar test experiments were performed to determine the optimal dose of aluminum sulfate. A set of experiments were performed using jar test to investigate the optimal dose of sludge to add in order to improve the quality of water.

The sludge produced by the treatment plant and used with the aluminum sulfate improves the turbidity removal percentage and achieve good removal efficiencies. The maximum value of turbidity removal at turbidity of 10 NTU is 98.41 corresponding to the optimum dose of sludge of 35 mg/l.

A set of experiments is performed using the optimal sludge dose obtained from the experiments below (35 mg/l) while the aluminum sulfate was changed from 0 to 10 mg/l. Figure 6 shows the effect of aluminum sulfate and sludge dose under various settling time on turbidity removal.

Table 4 gives the results of turbidity removal percentage using aluminum sulfate only, optimal dose of AS and sludge and the optimal dose of sludge and the proposed dose of AS.

The results show that the sludge used as coagulant aid with the aluminum sulfate improve not only the water quality produced by the treatment plant (turbidity removal percentage from 96.71% to 97.56% but also it can be used to reduce the aluminum sulfate dose in the coagulation process by 40%.

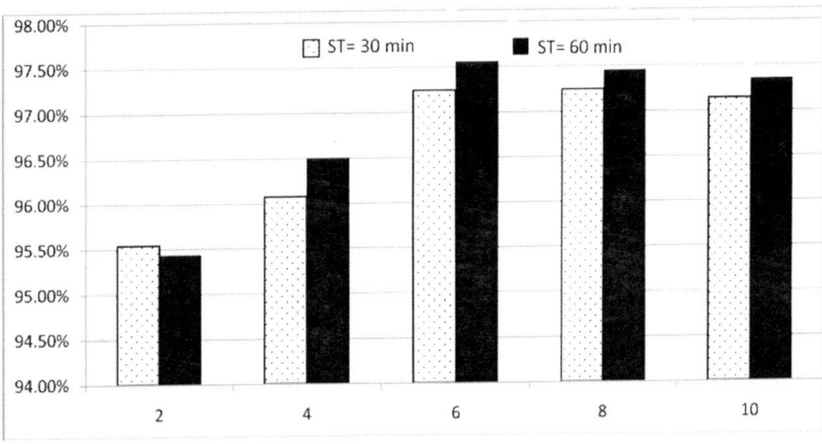

Figure 6. Effect of Aluminum sulfate doses (mg/l) and sludge dose (35 mg/l) under various settling time (ST) on turbidity removal percent (%).

Table 4. Turbidity removal percentage (%) using AS and sludge at settling time of 60 minutes.

	Turbidity = 10 NTU
Initial turbidity	10
Optimal dose of Aluminum sulfate (mg/l) (1)	10
Optimal dose of sludge (mg/l) used within AS (2)	35
Dose of AS (mg/l) proposed to be used with optimal dose of sludge (3)	6
Turbidity removal percentage using only AS (1)	96.71%
Turbidity removal percentage using AS and sludge as coagulant aid (1) + (2)	97.14%
Turbidity removal percentage using optimal dose of sludge and AS (2) + (3)	97.56%

4.3. Chlorination

The chlorine is the second chemical which impacts the cost of the treated water. The optimization of the consumption of this chemical can reduce the cost of the water produced by the water treatment plant. Some measures should be taken such as:

- On-line and laboratory instruments verification.
- Disinfection data handling and reporting.
- Verification of on-line chlorine residual analyzes.
- Verification of the performances of the on-line instruments with a grab sample measurement at least once every week.
- Respect of the manufacturer's instructions for instrument calibration.
- Completion of grab sample verification after changing reagents or doses or doing any other maintenance operations on the instrument.
- Incorporation of all manufacturer recommended instrument maintenance.
- Using a portable kit to verify performances of on-line instruments.
-

4.4. Energy

The energy consumed by the water treatment plant can be optimized by adopting many measures such as:

- Using of variables speed drives on pump motors.
- Lighting controls upgrade.
- The preventive maintenance of different equipment and motors.
- The replacement of existing pumps and motors by more efficient pumps and motors.

The implementation of such measures results in annual electric savings of more than 6% of the energy consumed by the water treatment plant.

5. CONCLUSION

This paper has presented some results concerning many practical methods used in the management of the water treatment plant, the use of sludge as coagulant aid with the aluminum sulfate in the water treatment plant. The sludge improves the quality of the produced water by increasing the coagulation efficiency and the aluminum sulfate dosage is decreased. The sludge volume is reduced and subsequently sludge management costs. The coagulation process is optimizing and this approach offers several significant economics and operational benefits such as it minimizes the aluminum dose required to 40%, reduces the sludge volume produced by the treatment plant and decreases the residual turbidity of the water. These advantages are very important not only for process economy through reducing the cost of treatment but also in the management of the sludge volume. Therefore, it is reasonable to consider this approach to be applied in the treatment plant for water with similar turbidity level. Also, the chlorine and energy consumption can be reduced in order to optimize the water treatment plant performances.

REFERENCES

1. Shen, Y.H. (2005) Treatment of Low Turbidity Water by Sweep Coagulation Using Bentonite. Journal of Chemical Technology and Biotechnology, 80, 581-586.
2. Mukheled, A. (2012) A Novel Water Pretreatment Approach for Turbidity Removal Using Date Seeds and Pollen Sheath. Journal of Water Resource and Protection, 4, 79-92.

3. Eman, N.A., Suleyman, A.M., Hamzah, M.S., Zahangir, Md.A. and Salleh, M.R.M. (2010) Production of Natural Coagulant from Moringa oleifera Seed for Application in Treatment of Low Turbidity Water. Journal of Water Resource and Protection, 2, 259-266.

4. Aho, I.M. and Lagasi, J.E. (2012) A New Water Treatment System Using Moringa oleifera Seed. American Journal of Scietific and Industrial Research, 3, 487-492.

5. Raghuwanshi, P.K., Mandloi, M., Sharma, A.J., Malviya, H.S. and Chaudhari, S. (2002) Improving Filtrate Quality Using Agrobased Materials as Coagulant Aids. Water Quality Research, 37, 745-756.

6. Sanghi, R., Bhatttacharya, B. and Singh, V. (2002) Cassia An-Gustifolia Seed Gum as an Effective Natural Coagulant for Decolourisation of Dye Solutions. Green Chemistry, 4, 252-254.

7. Diaz, A., Rincon, N., Escorihuela, A., Fernandez, N., Chacin, E. and Forster, C.F. (1999) A Preliminary Evaluation of Turbidity Removal by Natural Coagulants Indigenous to Venezuela. Process Biochemistry, 35, 391-395.

8. Sciban, M., Klasnja, M., Antov, M. and skrbic, B. (2009) Removal of Water Turbidity by Natural Coagulants Obtained from Chestnut and Acorn. Bioresource Technology, 100, 6639-6643.

9. Patale, V. and Parikh, P. (2010) A Preliminary Study on Coccinia indica Fruit Mucilage Extract as Coagulant-Flocculent for Turbid Water Treatment. Journal of Pure and Applied Sciences, 18, 27-30.

10. Sciban, M., Klasnja, M. and Stojimirovic, J. (2005) Investigation of Coagulation Activity of Natural Coagulants from Seeds of Different Leguminose Species. Acta Periodica Technologica, 36, 81-87.

11. Amitkumar, A.G. and Farooq, I.C. (2015) Performance Evaluation of Some Natural Coagulants. International Journal of Civil and Structural Engineering, 3, 368-375.

12. Nirmala Rani, C. and Jadhav, M.V. (2012) Enhancing Filtrate Quality of Turbid Water Incorporating Seeds of Strychnospotatorum, Pads of Cactus Opuntia and Mucilage Extracted from the Fruits of Coccinia indica as Coagulants. Journal of Environmental Research and Development, 7, 668-674.

13. Farhaoui, M. and Derraz, M. (2016) Optimizing Coagulation Process by Using Sludge Produced in the Water Treatment Plant. Journal of Chemical and Pharmaceutical Research, 8, 749-756.

14. Lind, C. (1994) Coagulation Control and Optimization: Part One. Public Works for October, 56-57.

15. Lind, C. (1994) Coagulation Control and Optimization: Part Two. Public Works for October, 32-33.

16. Liu, H.Y., Rong, Q.W., Xu, L.H. and Zhang, Y. (2013) Application of Automatic Control System in Water Plant in China. Journal of Convergence Information Technology (JCIT), 8, 243-250.

11

CHAPTER

Water Management in a State-Centered Environment: Water Governance Analysis of Uzbekistan

Munira Aminova [1,*] and Iskandar Abdullayev [2]

[1]*Faculty of Economics, Social and Political Sciences and the Solvay Business School, Vrije Universiteit Brussels, Pleinlaan 2, 1050 Brussels, Belgium*

[2]*Center for Development Research–ZEF, Department of Political and Cultural Change, Walter Flex Str. 3 Bonn, D-53113, Germany*

ABSTRACT

In the early 1990s the countries of Central Asia started their transformation to a market economy. Uzbekistan is in the throes of change and facing a huge restructuring task and a need to improve governance. Decades of central bureaucratic allocation of natural resources left national bureaucrats with little capacity to formulate their own sustainable policies. Lack of inclusive, long term oriented policies and mechanisms in management of common pool resources brought catastrophic results in the past. The study reveals that implanting a top-down quota policy has positive effects on a larger public. However it may become a threat to the long term environmental sustainability.

Keywords: water management; governance; stakeholder analysis; Central Asia; Uzbekistan

1. INTRODUCTION

1.1. Governance

Increasingly "nirvana" concepts such as integrated water resources, water governance [1] are penetrating into developing countries, creating a space for existence within the public domain. Supported and funded by International Funding Institutions (IFI) as part of Washington consensus [2] concepts such as good governance has become one of the most promoted ideas due to the increasing importance of all strata of society in matters which were so far function of a state.

Kaufman *et al.* [3] define *governance* as traditions and institutions that determine how authority is exercised in a particular country. This includes (1) the process by which governments are selected, held accountable, monitored, and replaced; (2) the capacity of governments to manage resources efficiently and formulate, implement, and enforce sound policies and regulations; and (3) the respect of citizens and the state for the institutions that govern economic and social interactions among them. *Good governance* refers to a strong and effective state that is accountable to its people, governed by a constitutional rule of law and able to provide a stable political environment [4]. Good governance is responsive, participatory, transparent, equitable, accountable, consensus oriented, effective and efficient and directed toward strategic vision [5].

Central Asian countries rank the lowest quintile on governance indicators among all countries of the world covered by the study of the World Bank [3]. One of the reasons is of course is the dramatic if not revolutionary transition of those countries in the early 1990s from planned to market based economy which affected the very foundations of their political, social and economic lives. Transition had deep implications on all branches of public life including on the natural resources management (NRM) policies of the Central Asian countries. The reform agenda was ambitious however; the development of public administration was far behind the economic reforms initiated. Low administrative capability gave rise to enormous economic and social costs. The old rules were outdated. New ones have not yet been drafted. Uzbekistan for instance, was trying to change incrementally. It was however '...deeply concerned with maintaining the capacity to govern in the face of great change' [6].

In the midst of reforms the government of Uzbekistan and many other transitional countries has realized the absence of frameworks, mechanisms, government personnel and resources to make the change happen. 'It became evident that the old regulatory framework and obsolete administrative structures were not capable of resolving the urgent and complicated problems of transition to a new political and economic reality' [7] since the government was still run by the same old 'nomenklatura' which was there before the transition. Situation is further complicated since 70 years of central planning made local civil servants 'handicap' in terms of policy making. The policies were formulated in Moscow and National Republics were to implement them only. So the capacity of individual republics to formulate strategy was limited due to historical reasons.

Uzbekistan for centuries has been characterised by traditions of hierarchy and authoritarianism among its settled population of the river valleys and oases. 'Authority is personalised and personal loyalties are deeply rooted' [8]. Even after 19 years of independence organisational structures are still hierarchical, this reinforces the command and control culture. One of the explanations given by scholars is that the hierarchy and authoritarian leadership style fits the nature of an irrigated oasis society, so participative decision making and inclusion of stakeholders into the process of policy formulation seems against the local ideology.

Local ideology or local culture which is a 'unique hybrid of national and Muslim symbolism' nurtured throughout centuries due to country's privileged position at the crossroads of the ancient trading routes. It is questionable which cultural grouping to fit Uzbekistan into—how 'Asian' is a country that for about a century was ruled from Moscow, only becoming independent in 1991. Likewise, while the country's population is predominantly Muslim, again the experience of being largely isolated from the larger Muslim world when part of the Soviet Union means that the impact of religious affiliation differs from that in other parts of the Muslim world [9]. Religious rituals or so called 'community practices' got modified to suit the needs of society. In three generation Uzbek family living in the same household deaths, marriages, child births and others are still celebrated with the direct participation of community members where unwritten rules are unquestionably followed by the members of society and authority and hierarchy of 'respected' people are subordinated. So the transferability of the western concepts of governance into the countries which are so deeply rooted in tradition and shaped by soviet system like Central Asian countries is a case in question. Thus it has far reaching impact on water governance issues too.

1.2. Water Governance

Good governance contributes to sustainable water management. The concept of governance applies to the water sector [10,11]. Water governance described by Rogers and Hall [12], as 'the range of political, social, economic and administrative systems that are in place to develop and manage water resources, and the delivery of water services, at different levels of society'. DFID describes water governance as 'encompassing all the mechanisms, processes, relationships and institutions through which citizens and groups articulate their interests and exercise their rights and obligations'. The representation of various interests in water decision making and the role of politics are important components in addressing governance dynamics [13]. So participation, transparency and accountability form the guiding principles needed for ensuring policies and decisions on water are responsive to citizens [14].

Water governance has become a centrepiece of high level political agenda of the last decade, e.g., in 2000, Hague Ministerial Declaration [15] called for 'governing water wisely to ensure good governance so that the involvement of the public and interests of all stakeholders are included in the management of water resources'. In Bonn Freshwater Conference [16], ministries have proposed that each country should take appropriate measures for ensuring good governance of water. United Nations Millennium Assembly [17] urged to stop unsustainable exploitation of water resources and to develop water management strategies for the regional, national and local levels on improving water governance. However, as Franks and Cleaver [18] state, although there is a growing recognition of the importance of water governance issues surprisingly there is a lack of theoretical analysis and debate over water governance. Extensive research has been carried out on developing of water governance analysis frameworks and their application for various countries; however, in case of former Soviet states the water governance analysis is rare or scattered in nature.

Water management during the Former Soviet Union (FSU) in Central Asia was a part of the overall state policy towards increasing agricultural production, specifically cotton production. The resources, e.g., water were part of the input system which was controlled by the government. The consumers, large farms utilized all resources provided by the state for growing cotton in the collective land with collective (ir-) responsibility. This has changed after the independence, the previously common water resources became contested at the higher—interstate level; land reforms have led to the formation of numerous individual farms instead of large collective farms. This has increased competition and contestation for water at the lowest—grass root levels [19]. For the analysis of the impact of the transition on water governance authors have selected the state quota policy which is a major issue impacting water management since soviet times. Although state-quota policy normally faces variety of criticisms by researchers [20] serious and systematic academic research on this topic is lacking except for few recent attempts [21,22]. Existing literature on water governance research on Central Asia is scattered in nature and touch different parts of water governance, e.g., emerging and changing institutions, corruption, path dependency of new institutions, organizational changes in water sector.

This paper starts illustrating the changes which has been taking place within the government of transition countries and then follows on reflecting on the changes taking place in agriculture management of Uzbekistan. More specifically the paper focuses on the policy process within the agriculture management and its impacts to the local population. It then makes in-depth analysis of state quota policy and its implication to the general water management situation in Uzbekistan and on various stakeholders. The authors believe that this initial research which combines institutional appraisal and social analysis brings out many issues which can not be studied solely and integrated research approach is required to see the interrelatedness of the conflicting opinions around one issue. Applying inter-disciplinary approach and illustrating the need for this we believe that more research will follow.

2. OVERVIEW

State Quota

Uzbekistan is the most populous country of Central Asia accounting for more than 45% of the population of the region. Agriculture is the backbone of Uzbekistan's economy, contributing almost to one-third to the annual GDP, 40% of total employment and 60% of foreign exchange earnings. It is also a great consumer of the natural resources, and it accounts for 92% of Uzbekistan's 56 billion cubic meters (BCM) total water use [23], equivalent to 60% of all water use in Central Asia [22].

Cotton was grown in Uzbekistan for centuries due to the favourable environmental conditions. Although cotton was a dominant product of Uzbek agriculture for a long time starting from 1950s its production grew dramatically when Uzbekistan was proclaimed as a centre of nation's cotton production. Soviet planners extended irrigation and supported the growth of the cotton by critical inputs like machinery, fertilisers and most importantly water. Cotton was then bartered with other republics in exchange for foodstuffs and other products. The production of inputs and outputs was strictly controlled; there was a quota on output and area, state purchase system and price, quantity of production and others; in 1991 the government still kept control of all production but the intention was to liberalise step-by-step as it was a policy for all sectors of economy.

After the independence and the collapse of centralised system which brought massive disruption and hardship to the lives of millions, it was decided to reconsider national food security and achieve grain (wheat) independence, since importing 3–4 million tons of wheat was a threat to national food security [20]. Large portion of land was allocated for growing wheat and by 1994 the wheat independence was achieved, this decreased the area for cotton growth to 30%–35%. Now cotton and wheat are two main agricultural products of Uzbekistan. Initially, the Uzbek government had quotas for almost all agricultural products. Since 1995, the quota system has been used only to cotton and wheat. Cotton production is controlled through quotas on area and output; similarly prices of inputs (such as fertilizers, diezel, *etc.*), outputs (such as cotton and wheat) and marketing (which are done by state companies to find appropriate markets for the sale of the outputs) are controlled too. In return farmers must sell 100% (only for cotton) of the output to the government for a fixed price. The quota system for wheat production is somewhat more flexible than that for cotton. Farmers are allowed to sell 50% of their quota in the market or to keep it for family consumption. The land to be sown with wheat is also strictly controlled and the same rules are applied as for cotton [24].

State quota for cotton and wheat means that the state through its semi-state organizations signs an agreement with individual farmers for delivering pre-determined amounts of wheat and cotton. On the basis of state quota farmers receive all necessary inputs to produce cotton or wheat. The farmers who are assigned by local governments (*hakimyat*) to grow cotton and wheat should follow strict rules of growing cotton-wheat, specifically: (1) agrotechnical rules (when to do sowing, irrigation, fertilization, and other.); (2) land use rules (where and how many hectares of which crop should be sown); (3) and administrative (attending daily meetings, obeying to state authorities, *etc.*). The plan for state quota is determined by local government (*hakimyat*) using land quality (bonitet) grades for each farmland). The state quota for cotton and wheat has the following important implications for water management: (i) water is seen as one of the inputs into the cotton and wheat quota production; therefore water management institutions (organizations, water rights, legal documents) are influenced by this fact, (ii) state quota allows administrative interferences into daily water management since it is a part of overall agriculture management [11]. The state quota shapes water governance process (because planning, distribution and control are done through state water management organizations) and since it is a part of the overall state policy, it pre-determines the water governance for irrigated agriculture.

The most malignant aspect of the cotton quota system is the allocation of specific areas for growing cotton, irrespective of their current appropriateness. Additionally even if farmers fulfill their cotton production quota, they still can not grow anything else in the area allocated for growing cotton. This gives farmers little incentive to increase land productivity and output since growing cotton is a lengthy and labor intensive process and if the right incentives does not exist there is a little hope that the efficiency of land will be increased.

3. RESEARCH FRAMEWORK AND METHODOLOGY

3.1. Water Governance Analysis

Dukhovniy *et al.* suggested 6 principle indicators to assess different water governance hierarchies in Central Asia (2008) [23]. The indicators are designed for the different water management hierarchical levels: (1) for transboundary level—is a minimum of unproductive loses; (2) for a national level—the contribution of water sector to GNP; (3) for sub basin level—the aggregated basin productivity, and so on. Although, the approach seems rather technical it reflects the important aspects of water governance which is missing in other methodologies.

Global Water Partnership's [12] water governance indicators are: openness and transparency, inclusiveness and communication, coherence and integration, equity and ethics, accountability, efficiency, responsiveness and sustainability. The centrepiece of governance analysis of GWP lies within IWRM principles, through which existing water governance deficiencies should be improved. However, application of the GWP approach is mainly limited to descriptive analysis of water sector.

According to the UNDP Water governance classification [11], the governance could be measured by analyzing following aspects of water governance: (i) equity, efficiency in water resources and services allocation, water administration based on catchments, the need for integrated water management approaches and the need to balance water use between socio-economic activities and ecosystems, (ii) clarification of the roles of government, civil society and the private sector and their responsibilities regarding ownership, management and administration of water resources and services, (iii) inter-sectoral dialogue and co-ordination , (iv) stakeholder participation and conflict resolution, (v) water rights and permits, (vi) the role of women in water management, (vii) water quantity and quality standards, (viii) bureaucratic obstacles and corruption, (ix) price regulation and subsidies and (x) tax incentives and credits. This approach although very comprehensive requires huge effort on collection of data on all those above aspects of water governance.

DFID's [24] formulated three indicators for good governance: state capability, accountability and responsiveness. The framework of DFID for governance analysis for CAR has been developed recently as an analytical tool to assess various dimensions of governance. The DFID CAR framework proposes the use of 12 indicators to assess governance. It fits well for the situations where state plays important role on provision of water services like Uzbekistan. However, extensive indicators makes application of the framework not only time consuming but results rather general.

Franks and Cleaver suggests looking into governance using four interlinked processes: resources for water governance, specific mechanisms of access, gendered outcomes for the poor and ecosystem outcomes. The framework of Franks and Cleaver is a useful tool to understand social dynamics of water governance. It helps to understand how different groups mobilize material and non material resources to produce variety of access mechanisms. It is also a useful tool for analysis of the impact of the different state policies on poor and ecosystems.

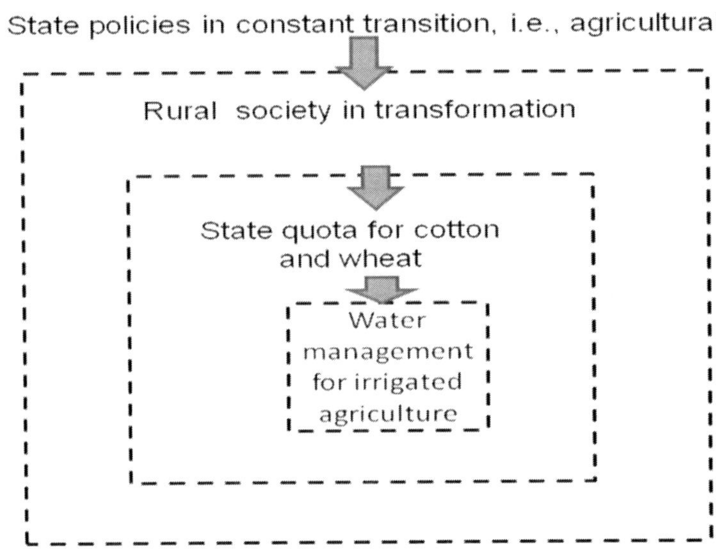

Figure 1. Water Management with wider context of transition.

Although there are comprehensive frameworks on water governance analysis, their application for the context of state-centric environments such as Uzbekistan is not an easy endeavor; water governance analysis for Central Asia region should consider particularities of the region's countries. The Central Asian states have been in transformation from centralized, planned socialistic state into some form of market economy since late 1980's. This transformation has been assessed as gradual [25], as failure and state-centric [26]. The water management in different states although went through different changes and transformations still remains quasi—state owned or managed at all levels. There are also different dynamics of the process at the different hierarchical levels. The water management exists within socio-economic and political systems where continuous changes are taking place, such as agricultural policies and those changes have a direct influence on water management (Figure 1).

A pre-requisite for governance analysis is the existence of minimum set of democratic institutions in the country. Application of any framework without adaptation into the Uzbek situation may result in rather vague, descriptive outcomes. The analysis of water governance should reflect how specific policy is made (policy process); who the actors are and what are their agendas (stakeholders); what are their mechanisms and resources to get access to the water resources (Table 1).

Table 1. A framework for water governance: major elements [18].

Major elements		Description
1. Process		Negotiation, decision-making and actions
2. Actors and agendas		Stakeholders in water governance
3. Resources		The generality of material and non-material resources in society
4. Mechanisms of access		Specific arrangements of resources sharing access to water
5. Outcomes	Societal outcomes	Positive or negative consequences for societies: on poor, gendered outcomes, and other.
	Ecosystem outcomes	Impact on resources and environment

Authors have selected a rather practical approach, combining Franks and Cleaver [18] framework for water governance and taking a case study approach to state quota policy using stakeholder analysis tool. Next section will further illustrate the details of the methodology used.

3.2. Analysis of Policy

Models and techniques in policy analysis are means by which arguments are clarified and the values and beliefs from which they are drawn are communicated effectively [22]. Policy analysis can be formally defined as 'an applied social science discipline which uses multiple methods of inquiry and arguments to produce and transform policy-relevant information that may be utilized in political settings to resolve policy problems [27]. We employed most existing techniques of policy analysis in our case study: those are discursive, interpretative, narrative and argumentative based approaches to policy analysis.

We used Geertz's 'thick description' [28] as a starting point to the ethnographic approach to participant observation. This approach takes a culture as a central point to describing the actions of actors involved, and gives meaning to those actions. It is both descriptive and interpretative. Beyond the basic description thick description makes the voices, feelings, actions and meanings of interacting individuals [29]. The advantage of the thick description in that it makes the agendas of elite and policy makers more transparent and accessible that maybe hidden deep within bureaucratic structure or behind a veil of scientific jargon [30]. We are trying to interpret the intentions and motives of different actors that are in fact forces that make things happen in real world of social action. Since in most cases those intentions are not written down or spelled out, it is our task as a narrative analyst to attach context specific common-sense meaning using the shared experience.

Interpretative approach emphasizes the treatment of differences as different ways of seeing, understanding, and doing based on different prior experiences [31]. *Narrative* scheme is believed to include the empirical causal connections between events but emphasizes social intentions and motivations. It is the understanding of the ways people are embedded in wider social context of situation and society [32]. Narratives create and shape social meaning by imposing a coherent interpretation of events and actions around us [33]. The narratives scheme of knowing configures the sequence of events and actions into a unified happening by identifying a larger pattern to which they contribute. It seeks to comprehend and convey direction and purpose to human affairs by knowing and explaining [34] and geared to the metaphor or contextualism [35]. *Stories* are inherently joint social productions [32], they are situated at the interface of the individual and his wider environment and are products of social groups in which they emerge.

We followed a process of policy metanarratives to examine the competing narratives in policy controversy introduced by Roe [36] in

1994. It follows a basic model with four steps. The first step is identifying the conventional and accepted narrative that dominated a policy controversy, specifically that: the state quota system negatively impacts production and efficiency and brings hardship to the lives of people involved. Second step is to look at the existence of other narratives (so called counterstories) that did not confirm to a dominant policy narrative. So in our case the counterstory is:the state quota policy does ***not*** onlynegatively impact production and efficiency and brings hardship to the lives of people involved. Then we took a third step which is to use a comparison of two narratives to 'tell' a metanarrative. Here we referred to the work of literary narration developed by Riffaterre [37], which seeks to account for two original policy narratives—each the polar opposite of the other—can both be the case at the same time, showing the ways that the same events can be retold from two different points of view [32]. By doing this we tried to ease out if not to remove the elements of controversy on the very sensitive issue of state quota policy and to move it to new grounds. And lastly we presented it in a comprehendible table of stakeholder analysis to make it open to empirical policy analysis for the given problem.

Roe's approach has been criticized for many shortcomings, which in our opinion does not reduce its usefulness [32]. We faced with two of the limitations. Although, both authors are researchers for many years, in one way or another both are beneficiaries of the given policy, this could have created a bias and the elements of social affiliation or political belief might have influenced our view of the current policy. Secondly, the process of constructing our metanarratives did not need to involve participants or actors of the policy controversy. This was largely our job as of policy analysts but the involvement of stakeholders could have produced useful insights into the process too.

3.3. Analysing Actors and Their Agendas

For analysis of state quota policy and its actors and agendas the most common used method would be Cost and Benefits analysis method (CBA). CBA identifies C&B of particular policy; screens the more important of the identified; enumerates each and compares the total costs and benefits, possibly by valuing them in money terms. This approach was also further improved by including secondary impacts analysis; intangibles like time and impact. Treatment of externalities, conventions about intangibles, time discounting approaches and using appropriate values other than money were further complications of this method [38]. Even after further improvements and additions CBA is criticised for favouring the rich and ignoring the politics of policy choice and the behaviour of winners and losers.

Cost-effectiveness analysis (CEA) assesses costs and their effect of different policy options. Best value is another term used in some countries like UK to represent CEA. *Risk-risk assessment*—is similar to CEA but it compares options to meet a set of objectives, but compares the risk of each option. *Multi-criteria analysis* avoids valuation of intangibles, it recognises that intangibles need to be identified and considered separately [39,40]. This approach could have been useful in analysing the state quota, but seemed to limit itself outside the notion of participation and inclusion that authors wanted to assess.

Generally most of the tools of economic analysis ignore the behaviour of individuals and the impact of the groups to individuals. Those tools also view individuals, largely, as economic units and assume that their overriding objective is to improve their well-being, or 'utility'.

Stakeholder analysis (SA) is the process of identifying policy's key stakeholders an assessment of their interests and the ways in which these interests affect project riskiness and viability. It was given preference over other methods because it draws from social analysis and institutional appraisal and also contributes to combining this data in a single framework; it helps to identify conflict of interests between various stakeholders; it helps to assess the appropriate type of participation by different stakeholders at various stages [41]. SA is found to be more applicable to analysing actors and different agendas of stakeholders. It potentially can be useful for policy making process too; the impact of the decision on different segments of the society can be understood and policy can be formulated for taking into account wider public interest. This framework provides a setting to comprehend perspectives of people who have potential interests in, or can influence the success or otherwise of policy, however, this tool is rarely used for policy analysis. SA is useful both for analysis of existing policies and for analysis of new ones before their implementation.

4. RESULTS

4.1. Process

The water management under state quota system is part of centralized and hierarchical planning and distribution of the water in the irrigated agriculture. The water planning follows strict hierarchical levels, starting from interstate level up to WUA level (Figure 2). In each level water allocation decisions are made based on selected criteria (e.g., area for growing planned cotton and wheat and other irrigated crops). The decisions are made on water allocation with involvement of water managers and decision makers (*hakimyat*—provincial and district government officials). Although other stakeholders are not present in water allocation process they apply different mechanisms to get access to the water resources.

The water allocation under the state quota system makes the process very hierarchical and bureaucratic. The process of preparing of the water allocation involves hundreds of staff of Ministry of Agriculture in the different levels (central office, UIS, BUIS). The water allocation plans are prepared for growing (*vegetatsya* in local) season and for fall-winter (ne-*vegetatsya*) periods.

However, the direct participation of the stakeholders other than water professionals and administrative managers (*hakimyat* in local) in the process of the water allocation is limited. The involvement of the water users and non-irrigation sectors in the process of water allocation are limited to the pilot activities [15]. Figure 2 below illustrates water allocation hierarchies under the state quota system.

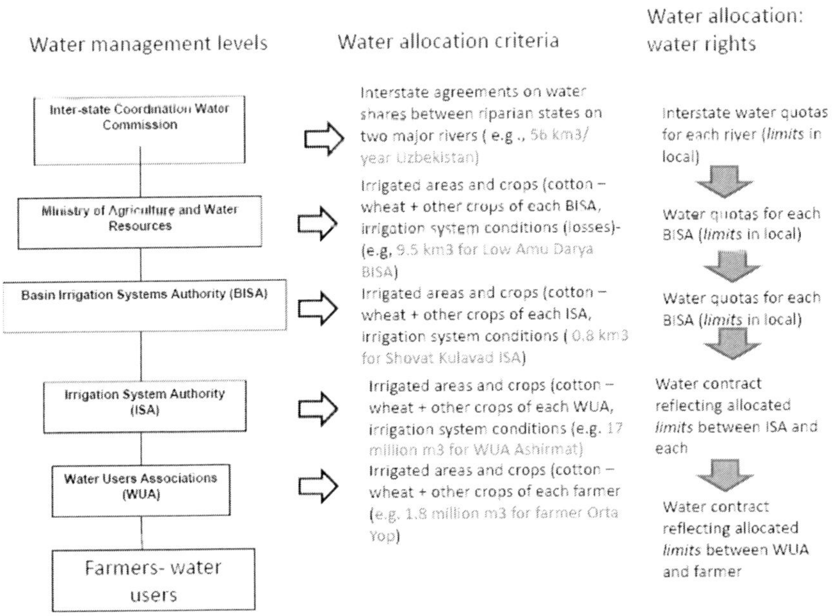

Figure 2. Water allocation hierarchy, criteria and water rights under state-quota system (e.g., Uzbekistan, Low Amu Amudarya BISA, one of ISA, WUA and farmer in Khorezm region).

4.2. Actors and Agendas (Stakeholders and Their Interests)

Context: Normally the participant observation is done by the external person who gets into the field and applies systematic rules and ethnographic algorithms to convey the meaning of the event back to the native. In our case the authors were parts of the culture in both frames of interpretation. We came to conclusions by observing the discourses around this topic. So from our experience two frames of interpretation came out:

- from population's point of view
- from the government's point of view

On one hand we find ourselves as a part of society, where we are unwillingly involved in everyday life and get information in different events involving local population, for instance, listening to taxi drivers' reflections when they give a ride (because of underemployment in the country people with higher education, formally engineers, lawyers can function as taxi drivers); through listening to neighbours in the street and relatives; by reading informal news sources, internet publications and news*etc*; during the debates with other academic colleagues who have more scientific approach to events *etc*.

Additionally, both authors formally had experience in being involved in the process of picking cotton, as a part of higher education system students are taken out to help farmers during harvest period. Normally with only 1 adult supervisor full time being responsible for 20–100 students (depending on the size of the course) of 17–21 years of age with superficial medical support and lack of basic hygiene. The daily norm for collecting cotton manually is set to be 60 kg (the norm of late 1990s). The hard physical labour of 8 hours per day with 1 hour break for lunch under the direct sun exposure was an experience which made us participant observers to the process of growing and picking cotton and a part of state quota policy. Later, the author was also involved in Red Cross project that dealt with environmental consequences of Aral Sea disaster which were mainly the consequence of poor water management due to increased water consumption for the growth of cotton—the policy outcome that was not accounted for, while designing the cotton policy. The author visited many tuberculosis hospitals and other hospitals for respiratory diseases in affected areas around Aral Sea region, being exposed to the negative externalities of a policy which affected the lives of millions.

On the other hand working for a number of years in public sector and having extended network of people who still work in high-to-medium level positions makes it possible to have alternative view on the issue; it leads to realisation of complexities of policy process. Informal talks with high level officials in unofficial meetings; informal meetings with friends who work in public sector, by analyzing formal media: the positive image of the process of cotton harvest, sense of regional competition for yields, the formal thanking of farmers for the hard work and others gives a possibility to have a 'counterstory' on the issue of state quota policy. Also reading (and in some cases writing or editing) high quality academic papers of various organisations like IWMI and others on the issue of water governance gave possibility to analyse written discourse. The accumulated knowledge and the context led to the understanding and the ability to be able to interpret the policy from insider perspective.

4.2.1. Identifying stakeholders

Stakeholder tables below will analyse primary, secondary and indirect stakeholder of the state quota policy. The particular attention will be placed on the affects of state quota system into water governance in the region. By doing stakeholder analysis we assigned subjective meaning to social events, however we transformed the fragmented information into meaningful, structured and organized whole. Stakeholder analysis gives us possibility to stress which elements of the whole is more meaningful and consequently we can draw attention of audience to those elements. It is a useful process through which authors transmit the information by interpreting them through available social, cultural and psychological principles in hand [32].

The process of drawing up Table 2 by ODA [41]: first we draw up a list of all potential stakeholders, and then we divide them into two main categories: primary and secondary stakeholders. Primary stakeholders are those people and groups ultimately affected by the policy. This includes intended beneficiaries or those negatively affected. Normally primary stakeholders are divided by gender, social or income classes, occupational or service user groups. Secondary stakeholders are intermediaries in the process of implementation of the policy. They can be divided into funding, implementing, monitoring and advocacy organisations, or simply governmental, NGO and private sector organisations. In most cases it will also be necessary to consider key individuals as specific stakeholders (in our case those are Hokims (governors) of regions). Also there may be some informal groups of people who will act as intermediaries, for example, politicians, and local leaders, respected persons with social or religious influence. ODA suggested a checklist for identifying stakeholders, with slight modification we have applied it to our policy analysis:

Checklist for identifying stakeholders:

- List all primary and secondary stakeholders
- Identify all potential supporters and opponents of the policy
- Identify different types of female stakeholder to take gender into account (at both primary and secondary levels)
- Divide primary stakeholders into user/occupational groups, or income groups
- Identify the interests of vulnerable groups (especially the poor)
- Analyse if there are any new primary or secondary stakeholders that are likely to emerge as a result of the policy

The resulting list of stakeholders forms the basis of a tabulation of each stakeholder's interests in the policy, and the likely impact of the policy on them. Interests of all types of stakeholders may be difficult to define, especially if they are "hidden", or in contradiction with the openly stated aims of the organisations or groups involved. A rule of thumb is to relate each stakeholder to the objectives of the policy [38]. The below checklist by ODA helps to draw out interests of stakeholder:

Checklist for drawing out interests

- What are the stakeholder's expectations of the policy?
- What benefits are there likely to be for the stakeholders?
- What resources will the stakeholder wish to commit (or avoid committing)?
- What other interests does the stakeholder have which may conflict with the policy?
- How does the stakeholder regard others in the list?

Table 2. Stakeholder Table for a State Quota Policy.

#	Primary Stakeholders	Interests	Likely impact of the policy	Relative priorities of interest
1	Local population (middle to low income groups)	*Protection from food deficit	(+)	1
		*Lower prices for agricultural products	(?)	
2	Farmers	*Freedom to grow crops they want	(–)	1
		*Subsidies inputs	(+)	
		*Higher profitability	(–)	
		*More employment opportunities	(–)	
		*Better working conditions	(?)	
		*Higher yields	(–)	
		*Expand production	(–)	
		*Labour efficiency	(–)	
		*Stability for planning and income	(+)	
3	The government	*Increase exports	(+)	1
		*Decrease imports	(+)	
		*Increase budget of the country by exporting cotton	(+)	
		*Reduce agricultural monopolies	(+)	
		*Achieve wheat independence	(+)	
	Secondary Stakeholders			
4	Ministry of Finance	*More inflows to the budget	(+)	2
		*Less expenditure	(+)	
5	Ministry of Agriculture and Water Management	*More control over agricultural production for better planning	(+)	2
		*Availability of more food for growing population	(?)	
		*Better water productivity	(–)	
		*Preservation of water resources	(–)	
6	Khokimiyats	*More control and more power	(+)	3
		*Less administrative burden	(–)	
7	Water Management Organizations	*Efficient use of water resources	(–)	2
		*Services paid on time for maintenance of irrigation systems	(–)	
8	Local *Khokims*/Mayors	*More power, more control	(+)	3
9	Children helping parents after school and students of HEI helping in cotton harvest	*No disruption of studies	(–)	5
#	**Indirect stakeholders**			
10	Dehkans	*Access to fertile land	(–)	4
		*Better access to water	(–)	
11	Banks and other financial institutions	*Return on investment	(?)	4
		*No bad credits		
		*Less subsidised credits		

12	Environmental organizations	*Preventing environmental degradation *Decrease salinization *Decrease soil erosion	(–) (?) (?)	5
13	People living in Aral Sea zone	*Protection from food deficit *Improved environmental situation	(–) (–)	4
14	Neighbouring counties	*Access and management of water in their territories	(?)	5

Note: + positive impact of the policy, – negative impact of the policy, ? unknown; 1—highest priority and 5—lowest priority.

Placing stakeholders into the table gives a good grasp of their multiple interests and the ways in which those interests affect the policy (positive, negative or unknown). Thus it combined both institutional appraisal and social analysis in a single framework. The stakeholders were divided into primary and secondary stakeholders in the above table, they were further classified according to their income classes, or type of the enterprises that are involved in (farmers, dehkhans). Those who participated in the delivery of the process, for instance funders Ministry of Finance; monitoring agencies like Hokimiyats; key individuals like Hokims and others were classified as secondary stakeholders.

The state quota policy is intended primarily to protect local population from food deficits; Uzbekistan is the most populous country among its neighbouring countries and accounts for over 45% of the total population of the region. Over 50% of the population of Uzbekistan is under 25 years of age which means it has a growing population and the demand for wheat will continue growing; at this point it seems quite necessary to have wheat independence. As we can see from the table the key stakeholders to which the state quota system is directed for are labelled 'local population' and the policy has a favourable impact on them.

Other primary stakeholders are farmers; they are in fact both primary stakeholders to whom the policy is intended for, since having a state to purchase their products, in theory, is a genuine support scheme that is used in most of the countries to protect local agricultural production. However, at the same time farmers are actually secondary stakeholders too, since they deal with the actual implementation of this policy. At the same time they are also a part of the middle-to-low income population group too, so there is a conflict of interest in relation to the policy within one stakeholder. So they are the ones who are affected most by this policy. However, they score negative on most of the interests listed. Their freedom to grow the desired crops is restricted by the policy; their desire to have higher profitability is also negative since the government buys their output on a fixed price which is much lower than in the international markets. They want to expand their production and use the land for growing of other products to which there is a higher demand in the local market, and they can't do it either. It is obvious that farmers are negatively impacted by this policy.

On the other hand there is a government [we refer to government as to a set of policy making institutions, that includes but not limited to—Ministry of Agriculture and Water Management, Cabinet of Ministers, Oliy Majlis (Parliament) and others] who is another stakeholder and we can see that the effects of the policy are mostly positive. The government has a possibility to increase export of cotton, decrease import of wheat, and achieve wheat independence.

We can also see that there is a conflict of interests between the key stakeholders: local population and farmers. Local population wants to have access to wheat and whereas farmers want to grow the crops they desire and thus better living incomes. In this case the policy is totally ignoring the latter and only focusing on the former.

4.2.2. Assessing 'Influence' and 'Importance' of stakeholders

In the table below we classified stakeholders according to their relative influence and importance to state quota policy. We here refer to influence as to how powerful a stakeholder is and we refer to importance as to those stakeholders whose problems, needs and interests are priority for the state quota policy. In the table below we see that section B lists and classifies stakeholders who have high degree of influence on the project and who also has high importance whereas in Section D the stakeholders have low importance and low influence. And within each section we further classify them according to the two criteria.

'Influence is the power which stakeholders have over a project—to control what decisions are made, facilitate its implementation, or exert influence which affects the project negatively. Influence is perhaps best understood as the extent to which people, groups or organisations (*i.e.*, stakeholders) are able to persuade or coerce others into making decisions, and following certain courses of action' [41].

Power may derive from the nature of a stakeholder's organisation, or their position in relation to other stakeholders (for example, line ministries which control budgets and other departments). Other forms of influence may be more informal (for example, personal connections to ruling politicians). It is also be necessary to consider stakeholders whose power, and therefore influence, will increase because of resources introduced by the policy [41].

Importance indicates the priority given to satisfying stakeholders' needs and interests through the policy. Importance is likely to be most obvious when stakeholder interests in a policy converge closely with government's objectives. 'Importance is distinct from influence. There will often be stakeholders, especially unorganised primary stakeholders, upon which the project places great priority (e.g., women, resource poor farmers, slum dwellers, ethnic minorities *etc.*). These stakeholders may have weak capacity to participate in the project, and limited power to influence key decisions' [41]. Some questions we considered while drafting a matrix:

- Which problems, affecting which stakeholders, does the policy seek to address or alleviate?
- For which stakeholders does the policy place a priority on meeting their needs, interests and expectations?
- Which stakeholder interests converge most closely with policy objectives?

Table 3. Matrix classification of stakeholders according to relative influence on, and importance on State Quota Policy, Uzbekistan.

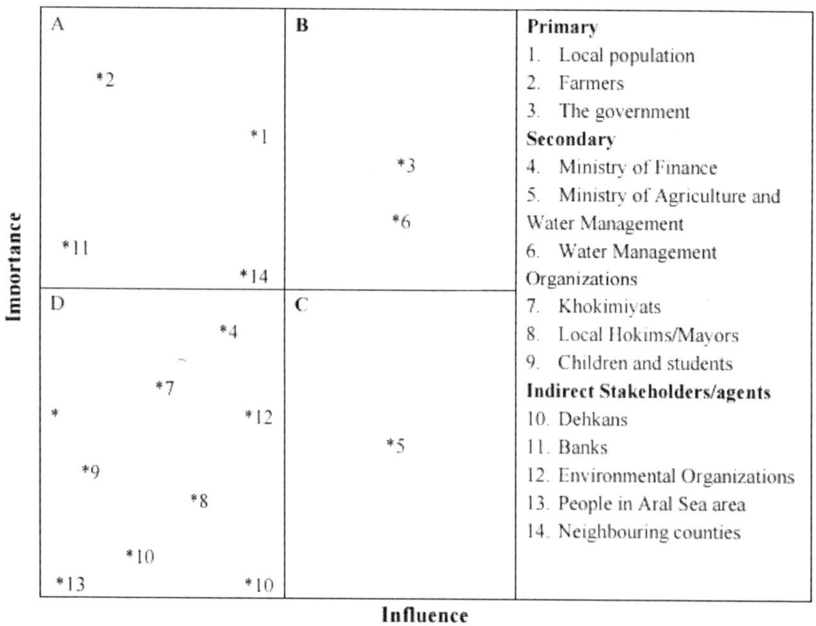

Legal hierarchy and political status give the Ministry of Finance and government the possibility to have a high degree of influence and high importance in our matrix. Whereas primary stakeholder like Farmers and Middle-to-low income population group falls into section A, since despite their importance they have little influence on the policy and they need special initiatives if their interests are to be protected. In box C there is stakeholder who has high influence but their interests are not in line with project objectives. In our block C we have a Ministry of Agriculture and Water Management.

This table matrix shows the power relations between the different stakeholders. We can observe that the large number of stakeholders fall into section D. with low importance and low influence. The matrix is another proof for the hierarchical nature of the society and that the power is accepted *de facto* in the society without questioning it and that the interests of low level population groups are not taken into account. If the numbers of low-income population grows, which is the case post-independence, without much social protection, hierarchical decision making which normally are made by those who do not fall into low-income level groups can create further social unrests in the society.

The above analysis gives an opportunity to assess major risks and negative responses on the policy. Normally for any policy to be successful its benefits should overweigh costs. In the case of state quota policy, in the first glance it seems to fulfil the stated aim which is to have a protection from food deficit and this benefit maybe much higher than many of the other social costs involved. Although in the long run if not taken into consideration the overall environmental situation may bring detrimental results for the local population. Normally when the policy is in the early stages of consideration risks and assumptions of it are considered relating to the primary stakeholders. It seems that in the case of state quota policy this has not been done or at least it is not as explicit as it should be.

4.2.3. Stakeholder participation

Stakeholder analysis contributes to the process of deciding how the key stakeholders are to be included in the policy process. "Key" stakeholders are those of high importance, high influence, or both. Key stakeholders with high influence and importance to project success are likely to provide the basis of the policy "coalition of support", and are potential partners in planning and implementation. Conversely, stakeholders with high influence, but with low importance to project success may be "managed" by being consulted or informed [41].

Social inclusion and participation is one of the key components of good governance; citizen participation is a cornerstone of democratic political processes [32], it is also one of the pre-requisites of improving water governance. If key stakeholders like local population and farmers are included in the initial stages of a policy and they participate in its design it will create the feeling of ownership for the project, objectives of the policy could serve as a binding idea to achieve as oppose to if a policy is simply imposed. 'By decreasing conflict and acceptance of or trust in decisions by the government agencies it can provide the citizens with an opportunity to learn about policy problems; such learning can improve the chances that they will support the resulting decisions' [32]. It also increases the chances of clearing up misunderstandings about the nature of controversy and the view of various participants. Stakeholder inclusion may take longer time but will generate commitment of those involved. The below table gives recommended scenario for stakeholder participation in the process of decision making and policy formulation concerning state quota policy. If all stakeholders listed are taken on board while making a decision or as early as agenda setting phase the results will likely be a better catered policy which takes into account needs of all strata of society.

Table 4 suggests a way of involving stakeholders in the design of the policy from early stages. It is clear that key stakeholders need to be at least informed and some with existing knowledge should rather be consulted from as early as identification stage. Some stakeholders like Dehkans who are affected by the outcomes of the policy should be included as partners during the entire process.

Inclusion of the stakeholders during the decision making makes policies better integrated with reality and normally leads to the design of policies catered for the needs of population from the point of view of public interests. It also enables the voice of the poor and most vulnerable to be heard. This will require the creation of additional mechanisms, methods and infrastructure to make it happen, however, will be a good start for building inclusive society and introducing elements of bottom-up policy. This will require on-going political will coupled with donor assistance.

However, hoping that participation of stakeholders will leave to a better policy outcome immediately would be politically naïve. Differences in power, social status, knowledge and other characteristics will play a role in making the voice of poor heard. So it is not a 'magic-cure' for economic and social problems. However, 'it does hold out the possibility of bringing forth new knowledge and ideas capable of creating and legitimizing new interests, reshaping our understanding of existing interests and influencing political pathways along which power and interest travels' [32]. Given the complexity of policy choices, decisions comprise a trade-off; there are normally some people who loose as a result of any policy and some who win. Therefore, ensuring that policy options are available to the public, and procedures exist to collect and comment upon public input, legitimizes policy decisions.

Table 4. Summary participation matrix for State Quota Policy.

Type of participation Stage in cycle	Inform	Consult	Partnership	Control
Identification	*Local population *Farmers	*Ministry of Agriculture and Water Management *Farmers *Water management organizations	*Ministry of Agriculture and Water Management	
Planning	*Local population *Farmers	*Ministry of Agriculture and Water Management *Farmers *Water management organizations *Ministry of Finance	*Environmental organizations *Dehkans	*Ministry of Finance
Implementation	*Farmers	*Ministry of Agriculture and Water Management *Farmers *Water management organizations	*Dehkans	*Ministry of Finance *Khokimiyats *Hokims
Monitoring and evaluation	*Farmers *Khokimiyats *Hokims	*Khokimiyats *Hokims	*Khokimiyats *Hokims	*Khokimiyats *Hokims

The idea of citizen participation is further complicated by local cultural pre-condition that authority is not to be questioned, and normally there is a very little tolerance for alternative views in Uzbekistan, if they are not coming from the people who are united around the same or at least a similar policy belief. These are normally the people who interact with each other on a regular basis, sharing more specific knowledge based understanding about problems and solutions. Normally those ideas which are snot in line with mainstream ideas on policy issues are not tolerated. So there is an issue of overcoming power distance within the society and inherent acceptance of hierarchy and the implication that inclusive decision making might mean weakness of the governor to governed, however, this issue is not discussed in-depth in the current paper; it is rather a topic for further research by the authors.

4.3. Resources

The different stakeholders use various techniques so called 'resources' to control the state quota. Following types of the resources are mainly utilised by stakeholders: (i) regulatory resources—access to decision making on water allocation, (ii) information resources—access to information regarding water allocation and distribution, (iii) professional resources—links with water management organizations (or individual bureaucrats in water sector), (iv) administrative resources—access to local government office (hakimyat) and (v) technological resources—access to technologies and infrastructure on water network. The application of the different resources depends on the position and agenda of the actor.

Agents	State	Khakim	WUA Chairman	Farmer	Rural people/villagers
Agendas	More production from a unit of area/ water	(province/district) state quota plan for his/her area fulfilled	water reaches his/her WUA and problem areas	all his/her fields are irrigated on time	water is enough for livelihood needs
Resources	Regulations, rules, state enforcement mechanisms, quota contracts	control of daily water distribution through WMO reps in rural areas to report and implement the concrete tasks, use of police to control implementation, use of personal networks in the locality	mobilization of staff, using links with khakim or other high level officials to get water to his/her area, mobilization of people-water users (both physical and monetary), appealing to local leaders, physical presence	Mobilization of his/her workers for irrigation, providing of support to WUAs, use of links, use of technology (pumps, pipes, tractors, *etc.*), participation and appeals to khakim	Appealing to rural leadership, WUA leader, staff, application of technologies, links, appealing to khakim, use of non—violent protest

Figure 3. Agents, agendas and resources in Uzbek setting.

If the state uses regulatory and administrative resources as the major mechanism to access water resources, the WUA chairman may apply professional and technological resources, whereas a farmer may apply technological and administrative resources, and others. The state-quota is a hierarchical process and therefore the access to different resources depends very much on where, in this hierarchical system, the actor is located. The higher the location of the actor within the hierarchy the better the access to different resources he/she has. This limits also weaker groups or individuals from the access to different resources including water resources and therefore leads to social exclusion.

4.4. Domains and Mechanisms of Access

The water distribution in comparison with allocation is more de-centralized process, because of huge irrigation network and many actors (stakeholders) involved in each level. Although national and regional level governments play an important role on water management issues they have no real means to distribute water. Because, water is actually distributed at the canal and Water Users Association (WUA) level, where mainly lower level water managers, e.g., canal managers and/or operators are controlling the gates and regulation infrastructure work on a daily basis. Therefore, although water allocation process is seen as very structured, hierarchical process its implementation (water distribution) often becomes daily 'water control' [42,43] struggle between different stakeholders (cotton-wheat farmers *vs.* other water users). Different players with different resources interact with each other in four domains with regard to access to the water resources: (i) administrative domain where local government lobby for state-quota interests, (ii) water management domain where water management organizations (WMOs) lobby for the interests of the state on water resources management, (iii) farmers lobby on their own behalf and (iv) farmers of different types interact with each other (as can be seen from the Figure 4 below). The de-collectivization has changed socio-political situations in the region and led to the formation of different interest groups. There are at least three different groups of farms in rural Uzbekistan [44]. The first group of farms are under state quota, growing cotton and wheat for the state with 10 to 20 ha of irrigated land, the second group is growing more commercial crops, such as rice, vegetables and fruits with land sizes around 1 ha and less. The third group is smallholder landowners which grow mainly crops for the subsistence of their livelihoods.

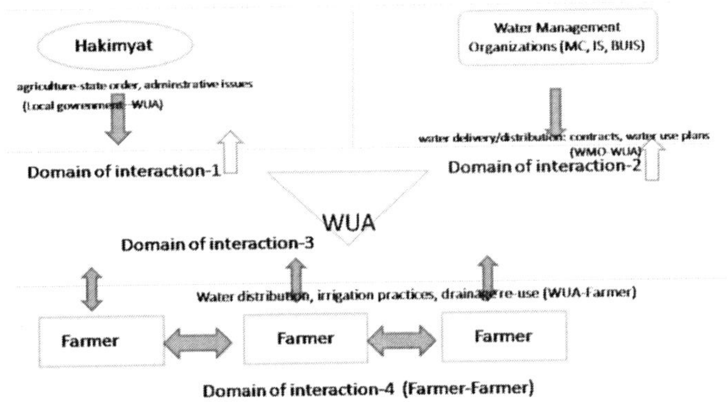

Figure 4. Domains of interactions on water management at the WUA level (Abdullaev forthcoming).

The different groups of farmers apply different mechanisms of access to the various resources. For example, the farmers with larger irrigated land and higher incomes buy diesel or electric pumps for irrigation of their fields. They do allow neighbour farmers and smallholders to use them, but only if the latter pay for the cost of operation of the pumps. The smaller land owners and weaker groups are not represented in WUA structures at present. WUA is also trying to change the existing situation for instance by assigning pumps to individual farmers. The operation and maintenance of the pumps, especially payments for electricity was a heavy burden on WUAs [19]. Although farmers under the state quota have priority on water distribution then other water users, the actual pattern of access depends on (i) their physical location within the irrigation network and (ii) their links within socio-political network. Most farmers under state quota system use their personal links with hakimyat, which is an administrative resource, as a last resource, in order to get access to irrigation water.

Access to water though, is not always directly dependent on Uzbek government. Central Asia has been considered as "a hot" area for international media and researchers for many years aftermath the collapse of the Soviet Union's [21]. Recently this issue again came back to the world public discussion arena. Failed to cooperate on transboundary water management, Central Asian states are staging again "cold war" on water issues. The regional bodies meant for transboundary water management are idle because of absence of political will for regional cooperation by riparian states. Arbiters on settling of the water conflicts are world powers Russia, EU, USA, and China, which are present in the region for a number of conflicting interests including its rich energy resources. Since the independence the large amounts of both grants and credits from international donors were provided for Central Asian countries to improve their water management and to build capacity of regional water institutions. For the last decade the interstate water institutions of Central Asia were considered as one of the best examples around world for developing viable annual water allocation agreements; for conducting regional cooperation on water research, data and information exchange; and capacity building [45,46]. However, at present interstate relations have worsened and interstate institutions alone cannot solve water problems of the region. The solution to the region's water problems lies not at the interstate level rather at the national water management systems of riparian states. CA states have excellent institutions at transboundary level; however, national water management institutions are creating obstacles for effective cooperation. Although there is no blueprint on the effective transboundary water management principles of integrated water resources management and river basin management approaches more specifically ownership and political commitment are good case for effective and equitable water allocation and harmonisation of water governance systems of riparian states.

4.5. Outcomes: Negative Externalities or Social Costs

The water management under state quota system is a complex process where state interests and interests of other stakeholders are integrated through bureaucratic water allocation and more chaotic water distribution processes. Differently from more open and democratic water governance, water management under cotton-quota system does not have feedback systems from different stakeholders. Therefore, actions of different stakeholders are not coordinated and regular clashes of their interests in the different domains of interactions take place. The most vulnerable are environment (nature) and smallholder water users. These two groups are disconnected from the use of various resources and access mechanisms to the water distribution.

The cotton-wheat quota system which requires extensive water resources already created a natural environment around irrigated areas. There are hundreds of lakes and swamps with diverse and rich biota. The lakes are fed by drainage water and excessive run off from irrigated areas. If water inflow to the irrigated areas will be reduced these lakes will start drying up, as a result causing similar problems as the Aral Sea. In most of the cases, the lakes are the source of fish, wildlife and income for local population. The other important issue which comes up as an environmental consequence of the irrigated agriculture is the dependence of as much as 60% of rural population for their livelihood uses. The local population in most of the irrigated rural villages are not provided with piped or safe drinking water sources therefore they use water from irrigation and drainage networks. The easiest way for them is either digging wells or hand pumps or getting water directly from the irrigation canals for livelihood uses. And at present, the above is not taken into account by existing water governance system. This may lead to the deterioration of environmental conditions, salinity, drying up of lakes, reduced biological diversity of the irrigated areas, lack of drinking water, and others. Therefore, the water usage in irrigated agriculture should be looked in the context of multi-purpose water use in the rural areas. The singling out only irrigation use will not reflect reality of water resources management in rural areas.

The emerging social and production differences in rural Central Asia as a result of land distribution have long-lasting impact on water management. The different groups use technical, organizational and socio-political 'water control' tools to get an access [47] for precious water resources. This dynamic situation on water management at grassroots levels have made water distribution more a social activity, the emerging new classes and differentiation in societies requires more careful planning of water management. The ignorance of this fact will lead to further deterioration of livelihoods of people who already were discriminated by de-collectivization. Continuing things as now and considering only state or semi-state bodies as major players of water management might lead to the collapse of existing mechanisms and institutions for water management [48].

5. DISCUSSIONS AND RESULTS

Lessons Learnt

1. Consultative and inclusive policy making and decentralisation of top-down policy making process and water decision making is a pre-requisite for formulating an effective and functioning policy. People in the grass roots, *i.e.,* those who are primary stakeholders like Farmers, Representatives of Dehkhan farms have an indigenous knowledge, the history and the tradition of managing agriculture and water in the region. For centuries these people have been growing cotton and wheat and other agricultural products in that area. Stakeholder participation may take longer time and effort and requires the set up of additional institutions or infrastructures. It would also requires a major shift in thinking of policy makers that the government is not a regulator any more, it is rather a 'service provider' and citizens are 'customers' of those services.

2. Implanting top-down quota system seems to have positive effects on a larger public and serves to the good intention of the government to have grain independence. However by allocating land for cotton and wheat the government is decreasing the motivation of farmers to get higher yields from the lands.

3. Western governance models might not be directly applicable to the countries like Uzbekistan in its entirety. Some adjustment is crucial for success. Donor led projects with heavy doses of external advice backed by resources is seen a norm for developing countries. By contrast western states are more autonomous in making decisions about their development paths [49]. With each new major shift in thinking practice in the west, there is first a period of attempted straightforward policy

transfer, followed by concern about failures of implementation, followed by an attempt to reform the government and institutions of developing countries [50]. It is also 'absence of informed analysis and the dominance of those concentrating on the transfer of the Western know-how that increase the danger of losing touch with the realities' [51].

4. The quota system although have many shortcomings has major role on regulating water management for irrigation. In other countries where market economy was supposed to regulate agriculture, water distribution ended up becoming a very chaotic process [52]. However, the system which is now in place ignores many emerging socio-political trends in the local level, creating space for conflicts in the future. Recognition of demands of non-irrigation users, *i.e.*, environment and livelihood needs into the water allocation process is a pre-requisite for sustainability of rural livelihoods. Improved water governance, such as inclusion of different stakeholders at least on water distribution decisions will reduce tensions over the water.

6. CONCLUSIONS

Socio-economic development is increasingly linked to vigour and effectiveness of policy making. Decision-makers recognise that without a reliable instrument for implementation, policies remain pipe dreams. In both, past and present, countries have struggled to create and manage public bureaucracies. From kings of the medieval era to the emperors of China, establishment and management of an efficient public administration has been on top of the agenda. If the example of Uzbekistan, its politics and transformation, so deeply rooted in tradition, so proud and self confident is to teach us a lesson, it is that autonomy in a modern period is limited and constrained, and a lot of effort and initiative is needed to translate recommendations and suggestions into co-ordinated policies and management systems.

The application of governance analysis framework for Uzbek conditions helped to shed a light on very complex socio-political process of water management under state quota system. The tool was instrumental for analysing water allocation and water distribution processes under state-centric environment. The identification of major stakeholders and their relevance to the state quota system also helped with identifying what resources, access mechanisms are needed for getting access to water. The lessons learnt from this research create an approach which could be applied for the water management governance analysis in similar conditions.

REFERENCES

1. Moll, F. Nirvana concepts, storylines and policy models: Insights from the water sector. *Water Alternatives* **2008**, *1*, 131–156.

2. Dellapenna, J.W. Adapting the law of water management to global climate change and other hydro political stresses. *J. Am. Water Resour. Assoc.* **1999**, *35*, 1301–1326.

3. Kaufmann, D.; Kray, A.; Mastruzzi, M. Governance matters VIII: aggregate and individual governance indicators, 1996–2008. In *Policy Research Working Paper Series, No. 4978*; Work Bank: Washington, DC, USA, 2009.

4. The World Bank. *World Development Report 1996: from Plan to Market*; Oxford University Press: New York, NY, USA, 1996.

5. *Central Asian Human Development Report: Bringing down the Barriers: Regional Cooperation for Human Development and Human Security*; United Nations Development Programme: New York, NY, USA, 2006.

6. Hesse, J.J. Rebuilding the state: public sector reform in Central and Eastern Europe. In *Public Sector Reform: Rationale, Trends and Problems*; Lane, J., Ed.; Sage Publications: London, UK, 1996; pp. 114–146.

7. Kotchegura, A. A decade of transition over: what is the administrative reform agenda. In *Civil Service Systems in Comparative Perspective*; Verheijen, T., Ed.; Edward Elgar: Cheltenham, UK, 1999; p. 21.

8. Gleason, G.; Motyl, A.J. *The Central Asian States*; Westview Press: Oxford, UK, 1997; p. 586.

9. Stevens, D.; Mukhamedova, L. Social responsibility at the grassroots: the influence of 'mahalla' community organisations on the CSR practices of small and medium enterprises (SMEs) in Uzbekistan. In *Culture and Corporate Governance*; Aras, G., Crowther, D., Eds.; Social Responsibility Research Network: Licester, UK, 2008.

10. Cosgrove, W.J.; Rijsberman, F.R. *World Water Vision: Making Water Everybody's Business*; Earthscan Publications: London, UK, 2000.

11. Abdullayev, I.; Kazbekov, J.; Manthritilake, H.; Jumaboev, K. Participatory water management at the main canal: a case from South Ferghana canal in Uzbekistan. *Agr. Water Manage.* **2009**, *96*, 317–329.

12. Rogers, P.; Hall, A. *Effective Water Governance*; TEC Background Papers No. 7; Global Water Partnership: Stockholm, Sweden, 2003.

13. UNDP Water Governance Facility. Available online: http://www.watergovernance.org/aboutwatergovernance/whatiswatergovernance.html (accessed on 15 November 2009).

14. *Building Integrity to Ensure Effective Water Governance*; Transparency International: Berlin, Germany, 2008.

15. Ministerial Declaration of The Hague on Water Security in the 21st Century. Available online: http://www.waternunc.com/gb/secwwf12.htm (accessed on 15 November 2009).

16. Bonn Freshwater Conference. Available online: http://www.africanwater.org/bonn.htm (accessed on 15 November 2009).

17. *United Nations Millennium Declaration*; United Nations: Geneva, Switzerland, 2000.

18. Frank, T.; Clever, F. Water governance and poverty: a framework for analysis. *Prog. Dev. Stud.* **2007**, *7*, 291–306.

19. Abdullayev, I.; Nurmetova, F.; Abdullaeva, F.; Lamers, J. Socio-technical aspects of water management in Uzbekistan: emerging water governance issues at the grass root level. In *Central Asian Water*; Rahaman, M., Varis, O., Eds.; Water & Development Publications: Helsinki, Finland, 2008; pp. 42–48.

20. Kandiyoti, D. Pathways of farm restructuring in Uzbekistan: pressures and outcomes. In *Transition, Institutions and the Rural Sector*; Spoor, M., Ed.; Lexington Books: Lanham, MD, USA, 2003; pp. 143–162.

21. Spoor, M. *Transition, Institutions, and the Rural Sector*; Lexington Books: Lanham, MD, USA, 2004.

22. Abdullayev, I.; de Fraiture, Ch.; Giordano, M.; Yakubov, M.; Rasulov, A. Agricultural water use and trade in Uzbekistan: situation and potential impacts of market liberalization. *Water Resour. Dev.* **2009**, *25*, 47–63.

23. Dukhovniy, V.; Mirzaev, N.; Sokolov, V. IWRM implementation: experiences with water sector reforms in Central Asia. In *Central Asian Water*; Rahaman, M., Varis, O., Eds.; Water & Development Publications: Helsinki, Finland, 2008; pp. 19–35.

24. Department for International Development. *Eliminating World Poverty: Making Governance Work for the Poor*; DFID: London, UK, 2006.

25. Spoor, M. *Globalisation, Poverty and Conflict: a Critical 'Development' Reader*; Kluwer Academic Publishers: Norwell, MA, USA, 1994.

26. Bloch, P.C. Agrarian reforms in Uzbekistan and other Central Asian countries. In *Land Tenure Center Working Paper Series*; Land Tenure Center, University of Wisconsin–Madison: Madison, WI, USA, 2002.

27. Dunn, W.N. *Public Policy Analysis*; Prentice-Hall: Englewood Cliffs, NJ, USA, 1981.

28. Geertz, C. *The Interpretation of Cultures*; Basic Books: New York, NY, USA, 1977.

29. Denzin, N. *Interpretative Interactionism*; Sage Publications: Newbury Park, CA, USA, 1989.

30. Thompson, W.B. Policy making through thick and thin: thick description as a methodology for communication and democracy. *Policy Sci.* **2001**, *24*, 63–77.

31. Yanow, D. The communication of policy meanings: implementation as interpretation and text.*Policy Sci.* **2000**, *26*, 41–66.

32. Fisher, F. *Reframing Public Policy: Discursive Politics and Deliberative Practices*; Oxford University Press: New York, NY, USA, 2003.

33. Kaplan, T.J. *The Narrative Structure of Policy Narratives: Beginnings, Middles, and Ends: the Argumentative Turn in Policy Analysis Planning*; Duke University Press: Durham, NC, USA, 1993.

34. Bruner, J. *Actual Minds Possible Worlds*; Harvard University Press: New York, NY, USA, 1986.

35. Sarbin, T.R. *Narrative Psychology: the Storied Nature of Human Conduct*; Praeger: New York, NY, USA, 1986.

36. Roe, E. *Narrative Policy Analysis*; Duke University Press: Durham, NC, USA, 1994.

37. Riffaterre, M. *Fictional Truth*; John Hopkins University Press: Baltimore, MD, USA, 1990.

38. Squire, L.; van der Tak, H. *Economic Analysis of Projects*; Johns Hopkins University Press: Baltimore, ML, USA, 1975.

39. Howlett, M.; Ramesh, M. *Studying Public Policy: Policy Cycles and Policy Subsystems*; Oxford University Press: Toronto, Canada, 1995.

40. Parsons, W. *Public Policy: an Introduction to the Theory and Practice of Policy Analysis*; Edward Elgar: Northampton, MA, USA, 1995.

41. *Guidance Note on How to Do Stakeholder Analysis of Aid Projects and Programmes*; Overseas Development Administration: London, UK, 1995.

42. Mollinga, P.P. Water policy—water politics: social engineering and strategic action in water sector reform. In *ZEF Working Paper Series 19*; Center for Development Research (ZEF): Bonn, Germany, 2009.

43. Mollinga, P.P. On the waterfront. Water distribution, technology and agrarian change in a South Indian canal irrigation system. In *ZEF Working Paper Series 19*; Center for Development Research (ZEF): Bonn, Germany, 2003.

44. Trevisani, T. Land and Power in Khorezm: Farmers, Communities and the State in Uzbekistan's Decollectivization Process. PhD Thesis, Institut für Ethnologie, Freie Universität Berlin, 2007.

45. Cai, X.M.; McKinney, D.C.; Rosegrant, M.W. Sustainability analysis for irrigation water management: concepts, methodology, and application to the Aral Sea region. In *EPTD Discussion Papers No. 86*; International Food Policy Research Institute (IFPRI): Washington, DC, USA, 2008.

46. Björn, G. Towards sustainable water management in Central Asia. In *Central Asian Water*; Rahaman, M., Varis, O., Eds.; Water & Development Publications: Helsinki, Finland, 2006.

47. Ribot, J.; Nancy, L.P. A theory of access. *Rural Sociol.* **2003**, *68*, 153–181.

48. Abdullayev, I.; Kazbekov, J.; Manthritilake, H.; Jumaboev, K. Participatory water management at the main canal: a case from South Ferghana canal in Uzbekistan. *Agr. Water Manage.* **2009**, *96*, 317–329.

49. Dolowitz, D.P.; Marsh, D. Who learns what from whom: a review of the policy transfer literature.*Polit. Stud.* **1996**, *44*, 343–357.

50. Batley, R.A.; Larbi, G.A. *The Changing Role of Government: The Reform of Public Services in Developing Countries*; Palgrave Macmillan: New York, NY, USA, 2004.

51. Hesse, J.J. Rebuilding the state: public sector reform in Central and Eastern Europe. In *Public Sector Reform: Rationale, Trends and Problems*; Lane, J., Ed.; Sage: London, UK, 1996; pp. 114–146.

52. Abdullaev, I.; Ul Hassan, M.; Manthrithilake, H.; Yakubov, M. *The Reliability Improvement in Irrigation Services: Application of Rotational Water Distribution to Tertiary Canals in Central Asia*; International Water Management Institution: Pelawatte, Sri Lanka, 2006.

Index